T0190496

Studies in Big Data

Volume 38

Series editor

Janusz Kacprzyk, Polish Academy of Sciences, Warsaw, Poland
e-mail: kacprzyk@ibspan.waw.pl

The series "Studies in Big Data" (SBD) publishes new developments and advances in the various areas of Big Data- quickly and with a high quality. The intent is to cover the theory, research, development, and applications of Big Data, as embedded in the fields of engineering, computer science, physics, economics and life sciences. The books of the series refer to the analysis and understanding of large, complex, and/or distributed data sets generated from recent digital sources coming from sensors or other physical instruments as well as simulations, crowd sourcing, social networks or other internet transactions, such as emails or video click streams and others. The series contains monographs, lecture notes and edited volumes in Big Data spanning the areas of computational intelligence including neural networks, evolutionary computation, soft computing, fuzzy systems, as well as artificial intelligence, data mining, modern statistics and operations research, as well as self-organizing systems. Of particular value to both the contributors and the readership are the short publication timeframe and the world-wide distribution, which enable both wide and rapid dissemination of research output.

More information about this series at http://www.springer.com/series/11970

Usha Mujoo Munshi · Neeta Verma
Editors

Data Science Landscape

Towards Research Standards and Protocols

 Springer

Editors
Usha Mujoo Munshi
Indian Institute of Public Administration
New Delhi, Delhi
India

Neeta Verma
National Informatics Centre
New Delhi, Delhi
India

ISSN 2197-6503   ISSN 2197-6511   (electronic)
Studies in Big Data
ISBN 978-981-13-3960-8   ISBN 978-981-10-7515-5   (eBook)
https://doi.org/10.1007/978-981-10-7515-5

© Springer Nature Singapore Pte Ltd. 2018
Softcover re-print of the Hardcover 1st edition 2018
This work is subject to copyright. All rights are reserved by the Publisher, whether the whole or part of the material is concerned, specifically the rights of translation, reprinting, reuse of illustrations, recitation, broadcasting, reproduction on microfilms or in any other physical way, and transmission or information storage and retrieval, electronic adaptation, computer software, or by similar or dissimilar methodology now known or hereafter developed.
The use of general descriptive names, registered names, trademarks, service marks, etc. in this publication does not imply, even in the absence of a specific statement, that such names are exempt from the relevant protective laws and regulations and therefore free for general use.
The publisher, the authors and the editors are safe to assume that the advice and information in this book are believed to be true and accurate at the date of publication. Neither the publisher nor the authors or the editors give a warranty, express or implied, with respect to the material contained herein or for any errors or omissions that may have been made. The publisher remains neutral with regard to jurisdictional claims in published maps and institutional affiliations.

Printed on acid-free paper

This Springer imprint is published by Springer Nature
The registered company is Springer Nature Singapore Pte Ltd.
The registered company address is: 152 Beach Road, #21-01/04 Gateway East, Singapore 189721, Singapore

Foreword

Decades back, the modern futurologist Alvin Toeffler stated that tomorrow knowledge will be power; that tomorrow is already with us. And Big Data is a significant source of knowledge generation. There are, of course, issues including security and privacy issues, in the generation of data and opening of Big Data in some areas. We also need Data Science standards and citation practices.

Data can be structured and also unstructured; it can be raw data or processed data. It can also be audio data or video images. We are already seeing an explosion in the generation of data, and this is only going to increase in future, with contributions also coming from the Internet of Things (IoT) and real-life applications driven by IoT devices. There is a need to develop algorithms, for storage, compilation, processing and analysis of such huge volumes of data. To understand this new domain, a new discipline of study has evolved as Data Science. It is felt that Data Science could be one of the most sought-after professions in the coming decades. There should, therefore, be a plan to introduce degree/diploma/certification courses in Data Science, which will foster skills and help employment generation in this field. It is also required to promote research in Data Science, by emphasizing its importance in our R&D value system.

Big and Open data are important for the growth of science. In fact, the scientists who create data have to be incentivized and supported through needed infrastructure, and their work cited and attributed with fairness. This is particularly needed if the data generated and processed are important for science or for society. To build any sustainable Open Data infrastructure, there is also the need for a robust policy framework. Policy provisions have a strong influence on strategy formulation and implementation of any Open Data programme. Therefore, a mechanism needs to be developed where these data can be stored, protected as well as archived for a longer time. Setting up Open Data infrastructure will make data a national asset, and researchers must share their knowledge in the "Open Data Repository" to create the collective knowledge-base. The use of Open Data generated by others would fasten the process of data collection and analysis and will save money and time. It would also help in improving India's overall standing in this research domain.

With a robust infrastructure, and utilizing developments in data storage, cloud computing technologies, data warehousing, and data mining, combined with AI and deep learning and machine learning techniques, data can be used for effective predictive modelling as well as for trending and for supporting government in decision-making and policy-making.

There is also the issue of quality of data. When I was a member of the Executive Committee of the "International Union of Crystallography" a couple of decades back, we were told that the Cambridge Organic Crystal Structure Database often found errors, particularly in the vibration amplitudes, derived from the anisotropic temperature factors, due to unknowingly mixing software using different formulae. Therefore, emphasis on the importance of reliable high-quality data and knowledge about the software used is necessary.

In India, we have a National Knowledge Network (NKN), for providing a unified high-speed optical fibre network which forms the backbone for all knowledge institutions in the country. The purpose of such a knowledge network lies in the country's quest for building quality institutions with necessary research facilities and creating a pool of highly trained professionals. NKN is designed as a Smart Ultra-High Bandwidth network that interconnects the leading knowledge institutions. NKN can play an important role in Big Data Science.

The availability of supercomputers, large data storage systems and the Internet provides opportunities, as never before, of manipulating, storing and accessing Big Data in Science and Technology. Our office has initiated synergizing the best practices in the Big Data experience of Astrophysics, Biology and Climate Science. The Data Centres would be located in the institutes with domain knowledge, with mirror sites located in the NKN Headquarters in New Delhi.

The present volume titled "Data Science Landscape: Towards Research Standards and Protocols" consisting of 24 chapters, written by eminent scholars from different parts of the world, is a timely publication. These chapters provide a current perspective of different areas on research and development, emphasizing the major challenging issues.

The volume concentrates on the important gaps in research data infrastructure, the absence of broad availability, use of data protocols and standards, and the like. The key issues of S&T, such as institutional, financial, sustainability, legal, IPR, data protocols, community norms and others, that need attention related to data management practices and protocols, coordinate area activities and promote common practices and standards of the research community globally are important, and some of these have been delved into this volume.

I congratulate the editors Dr. Usha Mujoo Munshi and Dr. Neeta Verma for the pioneering effort in bringing out this timely volume.

New Delhi, India   R. Chidambaram
October 2017   Principal Scientific Adviser
                                                  to the Government of India

Preface

Unprecedented explosion in the human capacity to acquire, store and manipulate data together with instant communication globally has transformed research from an era of data scarcity to data deluge, where experts see it as a "second revolution of discovery". Simultaneously, the growth of electronic publishing of the literature has created new challenges, such as the need for mechanisms for citing online references in ways that can assure discoverability and retrieval for many years into the future. Effective exploitation of "Big Data" basically depends on the international culture of "Open Data" involving data sharing, availability for reuse and repurposing. Therefore, there is a need to create the infrastructure, evolve methodologies, practices and most importantly policies that enable and empower researchers in identifying patterns and processes and subsequently analyse them to predict behaviour of complex systems, which so far had been beyond their capacity.

In order to foster push and pull for the Big Data applications in all segments of society and across disciplines, it is important to address multivariate issues that are critical for sharing, searchability and accessibility of data resources. For instance, Data Citation being one of the ways of giving attribution, its standards and good practices can form the basis for increased incentives, recognition and rewards for research data activities, that in many cases are currently lacking. Furthermore, the rapidly expanding universe of online digital data holds the promise of allowing peer examination and review of conclusions or analyses based on experimental or observational data, as well as the ability for subsequent users to make new and unforeseen uses and analyses of the same data—either in isolation, or in combination with other data sets. Accordingly, there is a need for strategy to be adapted to novel discoveries and approaches with the evolving needs of international research and the science community. Consequently, the crying need emerges for a framework of international agreements, practices or standards, national policies and practices for funding and incentivizing research. As questions of implications for stakeholders vary with disciplines, the diverse requirements of research communities in particular need to be addressed.

The idea of this work germinated from the two-day workshop on "Big and Open Data: Evolving Data Science Standards and Citation Attribution Practices", an international workshop which was attended by over 300 domain experts. The workshop focused on two priority areas: (i) Big and Open Data: Prioritizing, Addressing and Establishing Standards and Good Practices and (ii) Big and Open Data: Data Attribution and Citation Practices. This important international event was part of world initiative led by ICSU, CODATA-Data Citation Task Group.

The present edited volume deals with different contours of Data Science with special reference to data management for research innovation landscape. In all, there are 24 chapters written by eminent researchers in the field.

The issues concentrate on the important gap in research data infrastructure, the absence of broad availability- of the use of Data Citation protocols and the like. The funding agencies for research have begun to include data management plans as part of their selection and approval processes. The initiatives are already underway in different countries. The key issues of S&T that need attention broadly include, institutional, financial, sustainability, legal, IPR, data protocols, community norms and others, related to data management practices and protocols, and promoting common practices and standards globally for the research community are important.

The availability and application of data is of core importance in the data-centric world across disciplines right from science and technology to social sciences, arts and humanities to policy-making. Accessibility refers to the availability of domain-specific information to the user. In terms of generating applications, accessibility attributes the ease with which the existence of information can be ascertained, as well as the suitability of the form or medium through which the information can be accessed. The cost of the information may also be an aspect of accessibility for some users. With the explosion of social media sites and proliferation of digital computing devices and Internet access, massive amounts of public data are being generated on a daily basis. Efficient techniques/algorithms to analyse this massive amount of data can provide near real-time information about emerging trends and provide early warning in case of an imminent emergency. Careful mining of these data can reveal many useful indicators of socioeconomic and political events, which can help in establishing effective public policies for the purpose of human development. Thus, to decipher how can we harness the Big Data technologies to transform and revolutionize the developing world, we will have to consider pertinent issues like: How to access and use all of the data that are present out there on the isolated servers of various agencies and organizations for the development purposes? Where do we start and prioritize as to what particular areas of development can benefit from Big Data? Also, what are some of the well-known techniques for Big Data analytics that can be applied in multivariate contexts to benefit society at large?

In order to make the above happen, it becomes imperative to focus on setting up of robust infrastructure for storing, processing and analysing the data. Big Data has gained much attention from the academia and the IT industry alike. In the digital and computing world, information is generated and collected at a rate that rapidly exceeds the boundary range. As information is transferred and shared at light speed

on optic fibre and wireless networks, the volume of data and the speed of market growth increase. However, the fast growth rate of such large data generates numerous challenges, such as the rapid growth of data, transfer speed, diverse data and security. A vast majority of organizations spanning across industries are convinced of its usefulness, but the implementation focus is primarily application-oriented than infrastructure-oriented. However, the infrastructure architecture for any Big Data cluster is of critical importance because it affects the performance of the cluster. Modelling the infrastructure architecture for Big Data essentially requires balancing the cost and efficiency to meet the specific needs of businesses. While designing the Big Data architecture for the enterprise set-up, it is necessary to take a comprehensive approach such that the application requirements can drive the overall cluster design activity including cluster sizing, hardware architecture, network architecture, storage architecture and information security architecture. Therefore, the research directions need to facilitate the exploration of the domain and the development of optimal techniques to address Big Data to be optimally exploited for sustainable development and robust policy formulation.

Having focused on the need for robust infrastructure, the other aspects for discussion and requisite action include different dimensions of Data Science, such as right data quality, storage, metadata, accessibility, openness and interoperability.

Users have different expectations of data that are accurate and timely, comprehensive and cost-effective, locally relevant and also comparable with other similar situations as per the requirement—fit-for-purpose. Thus, it is of utmost importance to understand the definition of the so-called data quality, thereby outlining the various dimensions of quality and quality measurement, and have them integrated into quality assessment frameworks. For instance, typical examples of quality assessment frameworks include: European Statistical System (ESS), focusing on the statistical outputs and defining quality with reference to six criteria, IMF Data Quality Assessment Framework (DQAF), portraying holistic view of data quality, including governance of statistical system, and OECD Quality Measurement Framework, taking the user's side to approach quality and uses seven dimensions.

Metadata is essential for the interpretation of statistical data. Therefore, the attention to address various levels of metadata (e.g., structural, reference) is of prime importance. Together with metadata, the issues like openness and interoperability should be on the top of agenda to foster accessibility for reuse and repurposing of the data generated out of public funds.

While the promise of Big Data is real, there is currently a wide gap between its potential and its realization. A number of challenges lying in the pipeline are to be addressed in making an effective use of data discoverability and subsequent reuse. Heterogeneity, scale, timeliness, complexity and privacy problems with Big Data impede progress at all phases of the pipeline that can create value from data. The problems start right away during data acquisition, when the data tsunami requires us to make decisions, currently in an ad hoc manner, about what data to keep and what to discard, and how to store what we keep reliably with the right metadata. Much data today are not natively in a structured format. The value of data explodes when it can be linked with other data, thus, data integration is a major creator of value.

Since most data are directly generated in digital format today, we have the opportunity and the challenge both to influence the creation to facilitate later linkage and to automatically link previously created data. Data analysis, organization, retrieval and modelling are other foundational challenges. Data analysis is a clear bottleneck in many applications, both due to the lack of scalability of the underlying algorithms and due to the complexity of the data that needs to be analysed. Finally, presentation of the results and its interpretation by non-technical domain experts is crucial to extract actionable knowledge.

The sustainable business models around data repositories need to be pertinent to make the system of data archives as sustainable as text archives. The policy agreement is that the cost of data stewardship is an essential, integral part of the cost of doing research. Strong value proposition for data infrastructure and data sharing is required. Though, there are a few business models that are either largely structurally funded, or reliant on data access charges or membership fees, or exploring data deposit fees, or through substantial diversification such as propped up by project funding, or are supported by host institution. However, there is a need to do more work on the economics and business models of data infrastructure. Thus, there is a pressing need for work on who pays and how, analysis - of income streams, of innovative funding models, of willingness to pay and responsibilities and of business models in general.

In addition to the aspects touched above, the national/international perspectives of data and its various contours have also been portrayed through case studies in this volume.

The data are becoming pervasive in all spheres of human, economic and development activities. In this context, it is important for India to take a stock of what is being done in the data management area and begin to prioritize, consider and formulate adoption of a formal data management policy and procedures including a citation protocol for use by research communities in different disciplines and also address various technical research issues. The volume, thus, focuses on some of these issues drawing typical examples from various domains.

Thus, the volume covers interests of a wide range of readers including students, teachers, professionals, experts across domains and sectors as well as intellectuals interested in data game theory. Efforts have been made to minimize typos and other types of error.

Finally, it is our duty to acknowledge the technical committee of International Workshop on Big and Open Data: Evolving Data Science Standards and Citation Attribution Practices from where the idea of bringing this volume bloomed. We are also grateful to the management of Indian Institute of Public Administration (IIPA) and National Informatics Centre (NIC) for the encouragement in preparing the volume. We express our heartfelt gratitude to all the contributors of the chapters for their precious work that made it possible to bring out this volume.

New Delhi, India   Usha Mujoo Munshi
                                                                 Neeta Verma

Contents

Contents

About the Editors

Dr. Usha Mujoo Munshi is a Fulbright Scholar and is currently with Indian Institute of Public Administration (IIPA). Prior to joining IIPA, she has served in many prestigious organizations like INSDOC (Council of Scientific and Industrial Research), New Delhi; Indian National Science Academy, New Delhi and Indian Statistical Institute, Kolkata. She has over 100 research and conference publications and a few books to her credit, besides a number of other publications. Her publication christened "Multimedia Information Extraction and Digital Heritage Preservation" brought out by World Scientific is a detailed and distinctive anthology on various facets of the research issues in the area contributed by the world experts. Her publication on "Knowledge Management for Sustainable Development" brought out by Scientific International was released in October 2013 by the Vice President of India. Her latest publication on "Indian Higher Education—Research Landscape in the Global Context" by CBS Publishers and Distributors Pvt Ltd was released in 2016.

She is the recipient of several national and international awards that include Raizada Memorial Award 1994 for Young Information Scientist of the Society of

Information Science (SIS); SIS Fellowship in 1999; Fulbright Scholar in 1996–1997; ASSIST International Best Paper Award by ASSIST, USA, in 2002. In 2009, she obtained the post of DEA (directeur d'etudes associe), an academic recognition given to her by Fondation Maison des sciences de l'homme, Paris, France, and she is the first one to receive this recognition in the area of information science and technology.

She has organized over 50 national/international conferences and the latest being on "Big and Open Data: Evolving Data Science Standards & Citation Attribution Practices" held on 5–6 November 2015 at INSA, New Delhi.

She has substantially contributed to manpower development and organized a large number of training programmes on various themes.

She is on the editorial board of a few national/international journals in the area of information science and technology. She has been elected as the Editorial Board Member of the *Data Science Journal* which is dedicated to the advancement of data science. Acting as a member of various national and international committees, she was nominated as a member for the Steering Committee of Inter Academy Panel (IAP) and Task force on Digital Resources—US National Academy of Sciences, USA, and Co-Chair, Steering Committee, National Digital Preservation Programme (NDPP), Government of India. She has been nominated as member international "Data Policy Committee" by International Council for Science (ICSU)-CODATA, Paris. She has been associated with meetings of the Consultative Development Committee—set up by the Department of Science and Technology (DST), Government of India, for the Preparation of "Strategic Road Map for Data Analytics" for the government. She has been on the panel of International Scientific Programme Committee of the SciDataCon2016 of ICSU-CODATA, Paris; World Data System, Japan and Research Data Alliance. She is member of Task Group/s on Data At Risk (TG-DAR), and Data Citation (TG-DC), ICSU-CODATA. She has also been nominated as the Member, National CODATA Committee by INSA. Recently she has been elected as President, Information and Communication Society of India.

Neeta Verma is the Director General of National Informatics Centre, the premier ICT organization of Government of India implementing Digital India initiatives of the country. She has been associated with the introduction of World Wide Web Services in Government way back in 1996. She has consulted a large number of government departments on their portal development strategies.

As National Project Coordinator, she has led the Architecture, Design & Development of National Portal of India, india.gov.in, a gateway to Indian Government Information & Services in Cyber Space. She has also co-authored Guidelines for Indian Government Websites (GIGW), widely referred by Government Entities in India.

She has guided the team from concept to commission of first National Cloud for Government for electronic delivery of Citizen Services. She has led the implementation of Open Data in Government through Data Portal India, data.gov.in. She has also led the team in Design & Development of MyGov: the *largest* Crowd Sourcing and Collaboration Platform of Government of India.

Data Science Landscape: Tracking the Ecosystem

Usha Mujoo Munshi⬚

1 Introduction

Data is rapidly becoming pervasive in all fields. We have moved from data deficit to data deluge. The colossal data landscape can be attributed to the relentless parading technologies that are cropping up every now and then in the physical world. The smart devices such as sensors, smart energy meters, computers, smart phones, besides social media sites, multimedia content sites, trading and consumer data, individuals creating digital breadcrumbs and trails are constantly adding to the data landscape. The smart physical devices are in a position to sense, create, share, and communicate data across the globe in the era of Internet of Things (IoT). IoT is inter-networking of devices (physical, connected, smart) and has also been broadly defined as the infrastructure of the information society.[1,2,3] The enormous amounts of data available to science and society have been referred to as the fourth paradigm characterized by the data exploration, the first three science paradigms being empirical, theoretical, and computational branch.[4,5]

---

[1]Brown, Eric (13 September 2016). Who Needs the Internet of Things? Linux.com. Retrieved 20 April, 2017.
[2]Brown, Eric (20 September 2016). 21 Open Source Projects for IoT. Linux.com. Retrieved 27 March 2017.
[3]Internet of Things Global Standards Initiative. ITU. Retrieved 21 May 2017.
[4]Jim Gray on eScience: A transformed scientific method. Xvii–xxxxi. In The Fourth Paradigm: Data-intensive Scientific Discovery, edited by Tony Hey, Stewart Tansley and Kristin Tolle. USA, Microsoft Corporation, 2009. ISBN 978-0-9825442-0-4.
[5]Lynch, Clifford. Jim Gray's Fourth Paradigm and the Construction of the Scientific Record. In The Fourth Paradigm: Data-intensive Scientific Discovery, edited by Tony Hey, Stewart Tansley and Kristin Tolle. USA, Microsoft Corporation, 2009. ISBN 978-0-9825442-0-4.

U. M. Munshi (✉)
Indian Institute of Public Administration, New Delhi, India
e-mail: Umunshi.iipa@gov.in; umunshi@gmail.com

© Springer Nature Singapore Pte Ltd. 2018
U. M. Munshi and N. Verma (eds.), *Data Science Landscape*,
Studies in Big Data 38, https://doi.org/10.1007/978-981-10-7515-5_1

1

The entire research cycle from data capture to curation and to analysis and visualization contains large data chunks, but in the scholarly communications landscape, the published literature is a fraction of the data or what some say tip of data iceberg.[6] The volume of data captured does not necessarily get curated and leave aside actions regarding storage to searching and sharing to transfer. There have been instances that the data captured by the individual scientists remains in independent silos or on personal computers and even in some cases in files, left to dithering, volatile situation of data at risk for being lost forever or forgotten data in the know-how of authors only. Of late, endeavors are being made to address the issues of data at risk. For instance, with the CODATA Task Group, Data At Risk-Task Group (DAR-TG) set-up (late 2010, following the endorsement of the proposal for the DAR-TG at the CODATA biennial meeting in Stellenbosch, South Africa) primarily to bring to the forefront high potential of all heritage data and create avenues for data rescue efforts.[7] The DAR-TG members[8] took number of initiatives to address the basic issues such as Data-At-Risk Inventory (DARI) 2010–2013.[9]

Globally, there have since been a number of initiatives taken predominantly due to importance of data given both on research and governance front. The fact being that evidence-based policies in governance are required as against the intuition-based policies (which no longer work due to complex social structures), and also highly data-driven and data-centric research (for innovation) in a data-centric world is necessitated to unleash and unravel complex issues.

When we talk about the data-centric world, one essentially feels that it implies big data. The big data is not an unenviable phenomenon, where digital data crumbs continue pouring in at an unprecedented alacrity. But before we dissect the concept of "big data," one wonders whether today's semantics is going to stop at big or it can have hyped variants of the kind—large data, hyper data, bulky, outsized,... depending on the changing technological, economic, socio-cultural research ecosystem that sets the universal flavor. Empirical studies[10] also point on the changing usage of the term over the time. As few years down the road, big data seemingly was not a strong candidate to have created such hyped environment where data feels omnipresent even for activities and actions unimaginable. Therefore, it may be relevant to look closely at the bigger picture of health of data science landscape.

---

[6]Op cit. Jim Gray, xvii.

[7]http://www.codata.org/task-groups/data-at-risk.

[8]http://www.codata.org/task-groups/data-at-risk/dar-membership.

[9]http://cci.drexel.edu/mrc/projects/dari/.

[10]Matt Turck, Is big data still a thing. http://mattturck.com/big-data-landscape/.

2 Data Science Landscape

The data landscape has marked off from data deficit to data deluge (4Ds). This data deluge has great impact on the construction of the scientific record. The scientific record here represents research record which has been created in a systemic way. The research landscape is now witnessing data-centric and data-intensive research and science. This is hallmark of digitally enhanced research in progress (RIP), fostering a platform of knowledge-driven research infrastructure, leading to so-called machine-friendly scholarly communication system. Data has become a critical focus for scholarly communication, information management, and research policy. This unprecedented data tsunami offers major opportunities and profound challenges to research as perhaps never before, and experts see it as a second revolution of discovery.[11] The growth of electronic publishing of the literature has created new challenges, such as the need for mechanisms for citing online references in ways that can assure discoverability and retrieval for many years into the future. Till recently, data has been shelved mostly in independent silos. As such valuable data often is either not captured, cited, or re-used and re-purposed. The primary challenge therefore is to identify what part of these resources should be kept, the right ways to keep them, and the right tools and services to make them useful.

In the research publication system, we are very well versed in the paper publishing and literature archives, but our data publishing system has yet to be fully matured. Thus, the scholarly resource landscape covering articles, databases, and other literature resources were conveniently navigable and interoperable using the technological advancements for discovery and retrieval. But, when we talk of publishing data, nobody would argue that there has to be robust systems in place that not only holds data archives but can also address the issues of "data provenance" for proper re-use and re-purposing of the available data. In other words for the data at the unit level, we need to facilitate a researcher, how that number at unit level was arrived at and how he/she can re-use and re-purpose the same.

The sophisticated and people-friendly technological advancements are pouring and pushing data from all fronts and at all scales. The digital breadcrumbs or the bytes are though cropping up at high end while datasets are still at low end. The literature archives (journal papers and other research resources) have already matured; however, the digital data archives are also necessitated that have since started cropping up but yet to achieve full-scale system status. Though there are instances where people pickup datasets from various spaces and create another dataset, so mash-ups are happening. In the present context, where we see, feel, smell, and taste data everywhere, the interoperability of the data and literature archives is highly desired. The range of possibilities for a forward-looking approach where looking at metadata would facilitate researcher to hit the actual digital record

---

[11]Munshi, Usha Mujoo. International workshop on Big and Open Data—Evolving Data Science Standards & Citation Attribution Practices—A Report. Annual Report 2015–16, INSA, pp. 89–90.

and/or the associated data together with how to use (interpret, deduce meaning, and re-deploy or re-purpose) that data. For instance Entrez[12] search engine facilitates users to search several discrete health sciences databases[13] at the National Center for Biotechnology Information (NCBI). The user can look at an article, jump to gene data, follow the gene to the disease, and in fact go back and forth. It is an impressive case in point of the data and the literature (text) interoperating. This methodology is now beginning to spread, along with more generic forms of enhancement in knowledge representation facilitating auto-discovery and logical links to allied documents, data, actionable data within articles, and integration of data between articles. Such semantic enhancements in publications have started to be in vogue.[14,15]

The potential of sharing data for use/re-use/re-purposing is incredibly high and undeniably can transform the way the researcher works. Unless the data is re-used and re-purposed, the very meaning of the data archives is lost. There is definitely an upsurge in the digital data archives that is reflected in the impetus given to data management domain.

The experts have opined that "A true big data revolution should be one where data can be leveraged to change power structures and decision making processes, not just create insights".[16] While others[17] have referred to big data as tomorrow's knowledge having a great potential to lead the world. The data revolution that has led to the hyped term big data is going to change the status quo of the way we work, interpret, and deduce inferences, be it in the academic system/research sector/ government or the corporate arena. Having said this, before proceeding on to elaborate some of the issues regarding maturity model of the ecosystem, it is pertinent to muse over the basics of big data.

3   Big Data Paradigm

Tim Berners-Lee, inventor of the World Wide Web said, "Data is a precious thing and will last longer than the systems themselves".

---

[12]Entrez https://www.ncbi.nlm.nih.gov/books/NBK184582/.

[13]https://www.ncbi.nlm.nih.gov/Class/MLACourse/Original8Hour/Entrez/.

[14]www.rsc.org.

[15]Shotton, D. et al. Semantically enhanced version of a research article from PLoS Neglected Tropical Diseases http://purl.org/net/semanticpublication/Shotton_et_al_PLoS_enhancement_report.pdf.

[16]Emmanuel Lelouze. Big data for Development, 2012. UN Global Pulse White Paper. http://www.unglobalpulse.org/BigDataforDevWhitePaper.

[17]Chidambaram, R. In his inaugural address delivered during the International Workshop on "Big and open Data—Evolving data Science Standards and Citation Attribution Practices" held during November 5–6, 2015 at INSA, New Delhi.

The terms "data" and "information" are increasingly being used interchangeably, though there is a subtle difference between the two, which some may even argue. The data is to be treated as a source of information, and the processed or analyzed data leads to information.

The information explosion has always been thought of colossal information that comprises of collection of related data with context and perspective, and the data deluge shares though the same thought process, yet it is about mammoth raw data and hard facts, from which today's algorithms and powerful computers can reveal new insights that perhaps previously remained concealed. Thus, analytics of (big) data can lead to search for granular viewing patterns, providing powerfully easy paths to information and thereby impacting lives.

The big data, through data analytics, thus is supposed to portray a big picture about many a phenomenon hitherto unknown. The big data as defined by Danah Boyd and Kate Crawford is "a cultural, technological, and scholarly phenomenon that rests on the interplay of: (1) Technology: maximizing computation power and algorithmic accuracy to gather, analyze, link, and compare large datasets; (2) Analysis: drawing on large data sets to identify patterns in order to make economic, social, technical, and legal claims; (3) Mythology: the widespread belief that large data sets offer a higher form of intelligence and knowledge that can generate insights that were previously impossible, with the aura of truth, objectivity, and accuracy."[18]

The big data comprises the data environment in which the sheer volume of data is practically exploding each passing hour coming from all sorts of transactions including the real-time data as well. The big data landscape involves humongously large datasets that are characterized initially by *velocity* (where data comes at high speed on a real-time basis); *volume* (where data results into large files), and *variety* (where files come in various formats). Finally, *veracity* (credibility of the collected data) and *value*, where the inferences are drawn by computationally analyzing the data thus collected to reveal patterns, trends, etc. are other two characteristics of the Big Data. The other varied characteristics of the big data include the types of data that could be a combination of *structured* (number or facts, data in databases), *semi-structured* (smart devices leading to generation of data gathered remotely reflecting human actions), and *unstructured* (videos, blog posts, other social media content, documents). Most of the data in the big data landscape is increasingly unstructured and that is the real challenge to interpret for meaningful insights.

The veritable data analytics can generate evidence-based metrics, likely to drive innovative ideas and services. For instance through accurate analyses of vital information, one can observe release of socioeconomic value in business and government that on the other hand can foster responsible research and innovation in the scholarly research landscape.

---

[18]Boyd, Danah & Crawford, Kate (2012) Critical questions for Big Data, Information, Communication & Society, 15:5, 662–679, https://doi.org/10.1080/1369118X.2012.678878.

All the sectors, be it corporate houses like Amazon,[19,20,21] Google,[22,23,24] and IBM[25,26,27] are deploying, leveraging, and even offer tools to help sifting, and extracting data in the big data ecosystem. These tools facilitate in wielding big data for calculating return on investment (ROI), by developing workable strategies for gaining competitive edge. Similarly, the public sector such as healthcare, education, environment, and even banks[28,29] are gradually moving forward to integrate big data in their operations.

In the scholarly research landscape, effective exploitation of "big data" basically depends on the international culture of open data involving data sharing, availability for re-use and re-purposing. With increasing digital interactions, aggregating, analyzing, and re-purposing data to create new knowledge is likely to bring huge benefits in research landscape and other domains as well. Therefore, the real challenges are to create the infrastructure, evolve mechanisms and methodologies, practices and policies that enable researchers in identifying patterns and processes that so far had been beyond the capacity to resolve, analyze, and infer for predicting behavior of complex systems.

4 Maturing Ecosystem

Is data landscape including so-called big data ecosystem maturing? When we see torrent data cropping up from all the fronts, even from untailored conversations on the crowd sourcing platforms, the big question that crops up is about the preparedness in terms of the requisite infrastructure, methods, and methodologies to address challenges and avail opportunities to unleash the real potential of the data tsunami. The ecosystem thus needs sound policies, guidelines, standards, and protocols to address array of issues for making things happen that does not give insights but brings about paradigm shift in handling things in just and precise manner, thereby bringing about right kind of changes conducive for the sustainable

---

[19]Big Data Analytics Options on AWS https://d0.awsstatic.com/whitepapers/Big_Data_Analytics_Options_on_AWS.pdf. © 2016, Amazon Web Services, Inc. or its affiliates. All rights reserved.

[20]https://datafloq.com/read/amazon-leveraging-big-data/517.

[21]How Amazon Will Ride Big Data To $1 Trillion Market Cap https://www.forbes.com/sites/jonmarkman/2017/01/17/how-amazon-will-ride-big-data-to-1-trillion-marketcap/#437918f16333.

[22]https://datafloq.com/read/google-applies-big-data-infographic/385.

[23]https://cloud.google.com/solutions/big-data.

[24]https://www.quora.com/How-does-Google-use-big-data-to-give-effective-search-results.

[25]https://www.ibm.com/software/data/bigdata/.

[26]https://www-01.ibm.com/software/data/bigdata/business-partner.html.

[27]www.ibmbigdatahub.com/.

[28]Bholat, D. (2015), "Big Data and central banks", *Big Data & Society*, Vol. 1, pp. 1–6.

[29]Ng, V. (2012), "DBS bank implements big data analytics to enhance customer service", *Network World Asia*, Vol. 9 No. 3, pp. 28.

development. Therefore, it is pertinent here to examine some of these issues to comprehend the health of the ecosystem and how it continues to mature.

5 Data Ecosystem

A lot of literature have cropped up to focus on various contours of big data across sectors.[30,31,32,33,34,35,36,37,38,39,40,41,42] The data ecosystem is expected to provide an environment with cost-effective, innovative forms of information processing for enhanced insight and decision making by enabling tools, services, and workflows on the requisite infrastructure. The data driven innovations are unlocking great potential to tap new opportunities, provide new approach to research and development to move beyond the familiar one size fits all model. Government establishments including the higher education and research and development institutions are big data segments. A typical example could be that of Indian Power Grid (a big source of data, also big producer, 1.162 TB data archival per day), and it may be

[30]https://itif.org/publications/2016/02/29/europe-should-embrace-data-revolution.

[31]http://www.forbes.com/forbesinsights/sas/index.html.

[32]http://www.rss.org.uk/Images/PDF/influencing-change/2016/rss-report-opps-and-ethics-of-big-data-feb-2016.pdf.

[33]http://www.oreilly.com/data/free/archive.html#515.

[34]http://www.mckinsey.com/business-functions/mckinsey-digital/our-insights/digital-globalization-the-new-era-of-global-flows.

[35]https://www.forrester.com/report/TechRadar+Big+Data+Q1+2016/-/E-RES121460.

[36]http://www.forbes.com/sites/gilpress/2016/03/14/top-10-hot-big-data-tech.

[37]Cavanillas, Jose Maria, Curry, Edward, Wahlster, Wolfgang (Eds.). New Horizons for a Data-Driven Economy: A Roadmap for Usage and Exploitation of Big Data in Europe. Springer, 2016. ISBN: 978-3-319-21568-6 (Print) 978-3-319-21569-3 (Online) Available at: https://link.springer.com/book/10.1007%2F978-3-319-21569-3.

[38]Elston, Stephen F. Data Science in the Cloud with Microsoft Azure Machine Learning and Python. O'Reilly http://www.oreilly.com/data/free/data-science-microsoft-azure-ml-python.csp.

[39]Ping Chu, Li. Data Science for Modern Manufacturing: Global Trends: Big Data Analytics for the Industrial Internet of Things O'Reilly. http://www.oreilly.com/data/free/data-science-for-modern-manufacturing.csp.

[40]Informatica and Capgemini, The Big Data Payoff: Turning Big Data into Business Value. Report web version available at: https://www.capgemini.com/resource-file-access/resource/pdf/the_big_data_payoff_turning_big_data_into_business_value.pdf.

[41]Aureus Insights. Big Data & Analytics Trends—2016. http://blog.aureusanalytics.com/big-data-analyticstrends-2016/.

[42]Perspectives on Big Data, Ethics, and Society (Council for Big Data, Ethics, and Society).

indicated here that India is the largest operating synchronous grid in the world.[43,44] Nobody would argue that inclusion of technology made lots of data sources available. However, there are databases being generated by the government in administrative silos which need to be integrated. This integration necessitates straightening up of the processes and procedures such as classification, standards, structured data, or facilitating linked data, besides securing privacy and integrity.

In this multifarious data deluge environment, the big question is that how can establishments/agencies maintain buoyancy? A three-tier approach suggested by WIPRO[45] for managing big data that would be the key to stay tuned in this data environment. This tri-pronged approach includes—(i) the first tier to handle structured data, (ii) the second involving appliances for real-time processing, and (iii) the third for analyzing unstructured content.

This envisages that the key challenges lie in leveraging infrastructure for managing big data, data and metadata management, processing techniques and algorithms, integration of relentless parading technologies like IoT, cloud computing and the like, addressing real-time flow of data (data-on-the-fly) and other challenges concerning data security and privacy. Therefore, big data ecosystem broadly entails systems and processes for data generation, operationalization of the data (managing and storing, processing IT tools, and data preparation for analytics), data analytics including visualization, and finally data usage for business intelligence and decision support systems as well as linked data.

6 Data Policies, Standards, and Protocols

For digital globalization to be realized in the true spirit of the term, the necessity for making space for innovation to facilitate global data flows is the need of the hour. Therefore, transition from managing our stocks of knowledge to participating in flows, leveraging flows for learning, and accelerating capability building is necessitated. In order to make that happen, rhythm and rhyme of open access movement might help in sailing through. Just like the way, open access (OA) to scholarly communications was driven to success by various OA advocacy initiatives,[46,47] mandates, and other endeavors,[48] and in the same way it is important for

---

[43]http://timesofindia.indiatimes.com/india/All-India-Power-Engineers-Federation-Indian-power-system/articleshow/28294988.cms.

[44]http://www.livemint.com/Industry/KdUj7tUmfE0nCGCFfiS2eO/India-becomes-net-exporter-of-power-for-the-first-time.html.

[45]WIPRO. Big Data. http://www.wipro.com/documents/Big-Data.pdf.

[46]http://www.soros.org/openaccess/.

[47]http://www.digital-scholarship.org/oab/2statements.htm.

[48]http://www.unesco.org/new/en/communication-and-information/portals-and-platforms/goap/.

each country to have proper data frameworks in place to ensure the buoyant flow of data (through sharing, access, and use/re-use) for holistic development.

Establishing national policy and legal frameworks and data management practices for sharing, access, and re-use of data for re-purposing is crucial for all the countries in this data-centric world. The data-centric applications are now critical for any nation, government, or the organization.

Other initiatives regarding data management are burgeoning in the data-centric world of research and development across sectors and domains. The development of the reference model for big data is still[49] in progress. However, the 32N2386-Reference Model for big data[50] delineates guidance for big data consultation services and development solutions by covering whole processes from planning to discovery of big data.

International Organization for Standardization (ISO) has published[51] over 77 standards, and out of them over 26 are under development process[52] under the direct responsibility of ISO/IEC JTC 1/SC 32 working group (WG) on data management and interchange and ISO/IEC JTC1 SC32 WG2 for development and maintenance of various metadata standards.[53] Big data architecture models for management of big data systems[54] focus on six layers covering entire gamut of procedures from external data sources and sinks to applications and user interfaces.

There are evolving IEEE standards related to big data[55,56,57] for intercloud interoperability (IEEEP2302),[58] architectural framework for the Internet of Things (IoT) (IEEEP2413),[59] and analyzing reliability of data for equipment (IEEEP3006.8).[60] A number of international organizations like National Institute of Standards and Technology (NIST), Advancing Storage and Information Technology (SNIA), and Cloud Standards Customer Council (CSCC) are closely working on the standards and procedures for managing big data.

International Council for Science—Committee on Data for Science and Technology (ICSU-CODATA)—set-up International Data Policy Committee a few years down the road to provide inputs regarding the development and

---

[49]Big Data Reference Model https://bigdatawg.nist.gov/_uploadfiles/M0054_v1_8456980532.pdf.

[50]Jangwon Gim Sungjoon Lim Hanmin Jung. Next generation analytics and big data: ISO/IEC JTC1 SC32 Ad hoc meeting May 29, 2013, Gyeongju Korea. http://jtc1sc32.org/doc/N2351-2400/32N2386-Reference%20Model_for_BigData.pptx.pdf.

[51]https://www.iso.org/committee/45342.html.

[52]Ibid.

[53]http://metadata-standards.org/.

[54]http://bigdatawg.nist.gov/.

[55]https://standards.ieee.org/develop/project/2302.html.

[56]https://standards.ieee.org/develop/project/2413.html.

[57]http://standards.globalspec.com/std/1625530/ieee-3006-8.

[58]https://standards.ieee.org/develop/project/2302.html.

[59]https://standards.ieee.org/develop/project/2413.html.

[60]http://standards.globalspec.com/std/1625530/ieee-3006-8.

implementation of data policies to a range of international initiatives.[61] A report[62] submitted in May 2014 delved on the current best practice for research data management policies, which provides a framework for the development of data policies. CODATA has also constituted a task group who looks after coordinating data standards among Scientific Unions.[63] This is an extremely important move, since many organizations are engaged in creating standards, but whatever is created, many times are not visible to the majority and hence duplication is inevitable. So far as the data policy sources are concerned, there is quick guide to significant and useful documents relating to the development and implementation of research data policies.[64,65]

New emphasis on open access to data, increasingly important in this data-rich science era, is raising further issues with respect to security, privacy, ethics, intellectual property rights, formal publication of data, incentives to scientists/researchers and contributors to provide their data and other such issues that need earnest attention. Many nations world over have come out with national policies,[66,67,68,69,70,71,72,73,74] and addressed some of these critical issues.

In India, the "National Data Sharing and Access Policy (NDSAP)"[75] of Government of India launched in 2012 is the one accepted by all stakeholder groups regarding the required basic guidelines to follow for their projects. However, the policy (NDSAP) principles need to be wrapped with a continuous improvement loop as insights are fed back to improve the data analysis models as well as the process. These in turn can only drive the quality and completeness of data to further press forward future analysis. Another milestone in this direction has been the notification of the

---

[61]http://www.codata.org/strategic-initiatives/international-data-policy-committee.

[62]https://doi.org/10.5281/zenodo.27872.

[63]http://www.codata.org/task-groups/coordinating-data-standards.

[64]Research Councils UK Guidance on best practice in the management of research data: http://www.rcuk.ac.uk/RCUK-prod/assets/documents/documents/RCUKCommonPrinciplesonDataPolicy.pdf.

[65]DCC Guide: Five Steps to Developing an RDM Policy: http://www.dcc.ac.uk/sites/default/files/documents/publications/DCC-FiveStepsToDevelopingAnRDMpolicy.pdf.

[66]Tim Davies, Open Data Policies and Practice: An International Comparison, 2014 Available at: https://ecpr.eu/Filestore/PaperProposal/d591e267-cbee-4d5d-b699-7d0bda633e2e.pdf.

[67]https://opengovdata.io/2014/us-federal-open-data-policy/.

[68]http://www.rcuk.ac.uk/research/datapolicy/.

[69]https://ec.europa.eu/jrc/en/about/jrc-in-brief/data-policy.

[70]https://data.gov.in/opendatasites/denmark.

[71]www.government.se/about-the-website/psi-data/.

[72]https://opendata.riik.ee/en/roheline-raamat.

[73]www.gouvernement.fr/en/public-data-policy.

[74]www.opendatacharter.net/resource/g8-open-data-charter/.

[75]http://dst.gov.in/national-data-sharing-and-accessibility-policy-0.

Government Open Data License-India (https://data.gov.in/government-open-data-license-india) for specification of the data sets published under NDSAP and through the Open Government Data (OGD) Platform.

Some other policy-related notable initiatives to make huge impact on the socioeconomic landscape of the country include—The Federal Big Data Research and Development Strategic Plan[76] of the National Science and Technology Council (NSTC), prepared by the subcommittee on the Networking and Information Technology Research and Development Program (NITRD), and projecting the shared vision of the R&D. BioSharing (http://www.biosharing.org) a curated searchable portal of three registries of (over 600 community developed) content standards, (over 700) databases, and data policies in life sciences is yet another important example. The RDA Working Group (WG) on BioSharing Registry[77] and the outputs[78] of this WG include a set of recommendations and a curated registry providing cross-searchable records on content standards, databases, and data policies. Biodiversity Information Standards are also known as Taxonomic Databases Working Group (TDWG). TDWG Biodiversity Information Standards[79] focus on the development of standards for the exchange of biological/biodiversity data. There are host of standards[80] available in the life science field. Similarly, Global Earth Observation System of Systems (GEOSS) 10-year implementation plan for information flow and availability of data[81] and Public Library of Science (PLOS) Data Policy recommendation[82] wherein PLOS has identified a set of established repositories and data standards, which are recognized and trusted within their respective communities. There is a wide array of standards and databases mentioned in this policy.[83] In developing policies, laws, and practices at the national level, guidance can be found in the Organization for Economic Cooperation and Development's (OECD's) statements on access to publicly funded research data,[84]

---

[76]The Federal Big Data Research and Development Strategic Plan. Networking and Information Technology Research and Development Program (NITRD), May 2016. Available at: https://bigdatawg.nist.gov/pdf/bigdatardstrategicplan.pdf.

[77]BioSharing Registry: connecting data policies, standards & databases in life sciences WG https://www.rd-alliance.org/group/biosharing-registry-connecting-data-policies-standards-databases-life-sciences.html.

[78]Report detailing the WG outputs Available at: https://doi.org/10.15497/RDA00017.

[79]http://biosharing.org/collection/TDWGBiodiversity.

[80]https://biosharing.org/standards/.

[81]Group on Earth Observations (GEO), "GEOSS 10-Year Implementation Plan," adopted Feb. 16, 2005, p. 4, www.earthobservations.org/docs/10-Year%20Implementation%20Plan.pdf.

[82]https://biosharing.org/recommendation/PLOS.

[83]https://biosharing.org/bsg-p000036.

[84]OECD Principles and Guidelines for Access to Research Data from Public Funding http://www.oecd.org/science/sci-tech/oecdprinciplesandguidelinesforaccesstoresearchdatafrompublicfunding.htm.

the United States Office of Management and Budget's (US OMB's) Circular A-130,[85,86] and various European Union (EU) directives.[87,88,89]

In order to break barriers and make way for permitting research data flow afloat, creating a standard ecosystem is obligatory that would put in place a framework for critical issues like interoperability, taxonomies, security and privacy, reference architecture interfaces, use case, and the like in the big data milieu. Thus, for the suitable flow of global data and for global eResearch collaborations because of their complex behavior would necessitate across regions, appropriate national policy and legal frameworks, and data management practices to fulfill the commitment to data access and sharing for re-use.

7 Data Management Policies, Procedures, and Practices

There is a great deal of knowledge about data storage, data management, and access in the scientific community, and it is highly desirable to factor similar techniques in social and governance domains. Besides accrualling of the national data policies, the serious dialogues regarding data management plans (DMPs) have also started to ride waves.

The research data landscape demands well laid down process and framework for data management planning which includes (a) what you do with data during and after research endeavor and (b) how are you going to document the data for future use/re-use. With proper planning, many data management issues can be handled easily or avoided. There are general guidelines for data management[90,91] to facilitate

[85]Office of Management and Budget Circular A-130 on Management of Federal Information Resources (OMB Circular A-130), 2000, www.whitehouse.gov/omb/circulars/a130/a130trans4.html.

[86]Office of Management and Budget Circular A-16 on the Coordination of Geographic Information and Related Spatial Data Activities (OMB Circular A-16), issued Jan. 16, 1953, revised 1967, 1990, 2002, Sec. 8, www.whitehouse.gov/omb/circulars_a016_rev/#8.

[87]European Parliament and Council of the European Union, Directive 2003/98/EC of the European Parliament and of the Council of 17 November 2003 on the re-use of the public sector information, 2003, OJ L 345/90, http://eur-lex.europa.eu/LexUriServ/LexUriServ.do?uri=CELEX:32003L0098:EN:HTML.

[88]European Parliament and Council of the European Union, Directive 2007/2/EC of the European Parliament and of the Council of 14 March 2007 establishing an infrastructure for spatial information, 2007, OJ L 108/1, Apr. 25, 2007, http://eur-lex.europa.eu/LexUriServ/LexUriServ.do?uri=OJ:L:2007:108:0001:01:EN:HTML.

[89]European Parliament and Council of the European Union, Directive 2003/4/EC of the European Parliament and of the Council of 28 January 2003 on public access to environmental information and Repealing Council Directive 90/313/EEC OJL 041, Feb. 14, 2003, pp. 0026–0032.

[90]https://dmptool.org/dm_guidance.

[91]EUROPEAN COMMISSION Directorate-General for Research & Innovation. H2020 Programme Guidelines on FAIR Data Management in Horizon 2020 http://ec.europa.eu/research/

creating data management plan (DMP). The need for creating DMP has become more widely desirable because of the emphasis given both by the funders, (funding extramural research/researchers) and institutions (for intramural researchers) to develop data management plans for describing long-term preservation and access to data. Since the activities around DMPs are incessantly evolving, hence it sounds natural to learn about the stakeholder experiences with the specific templates, for instance Open AIRE survey about Horizon2020 template for DMPs.[92] It may be worthwhile to indicate that the DMPTool[93] as a one-stop-shop site facilitates in providing general management guidance, creation of DMP, access to resources, and the like. The information regarding DMP templates[94] both institutional and public ones are also available. The Research Data Alliance (RDA) has facilitated free research data management plan toolkit[95] to support the implementation process. The CODATA-RDA Legal Interoperability Interest Group[96] has studied the issues related to the intellectual property of data, and the resulting outcome is a set of high-level guiding principles and practical implementation guidelines[97] for all stakeholders. The recent efforts of the International Research Data Management (IRiDiuM) Glossary[98] and CASRAI Dictionary[99] have been designed to provide practical reference to all stakeholders associated with the improvement of research data management and the discussion thereof. Principally, creating standard glossary of community accepted terms and definitions, persevering it as a living document by continually updating it, and developing standard dictionary of research administration information meant to serve as a key component of interoperability strategies of research organizations are of prime importance.

Several funding bodies which mandate open access also mandate open data. National Science Foundation has already mandated submission of DMP for dissemination and sharing of the research results for re-use along with funding proposals.[100] Other funders, such as the European Commission (H2020 Programme Guidelines to the Rules on Open Access to Scientific Publications and Open Access

---

participants/data/ref/h2020/grants_manual/hi/oa_pilot/h2020-hi-oa-data-mgt_en.pdf (A template for writing a Data Management Plan (DMP) is provided in the annex of those Guidelines).

[92]OpenAIRE survey about Horizon2020 template for Data Management Plans https://www.surveymonkey.com/r/OpenAIRE_DMP_survey.

[93]https://dmptool.org/.

[94]https://dmptool.org/guidance.

[95]https://www.rd-alliance.org/free-research-data-management-toolkit-learn-project-now-available-download.

[96]http://www.codata.org/working-groups/legal-interoperability.

[97]https://zenodo.org/record/162241.

[98]http://www.codata.org/working-groups/standard-glossary-for-research-data-management-iridium.

[99]http://dictionary.casrai.org/Category:Research_Data_Domain.

[100]https://www.nsf.gov/bfa/dias/policy/dmp.jsp.

to Research Data in Horizon 2020),[101,102] the Wellcome Trust,[103] the National Institute of Health (NIH Sharing Policies and Related Guidance on NIH-Funded Research Resources),[104] Research Councils UK,[105] and many more are now asking researchers to include data management plans for sharing and using their data. Although many see the move to foster data availability and accessibility attributed to the imperative part of the May 2016 recommendations[106] from the European Council in response to a draft proposal by the Netherlands EU Presidency (Amsterdam Call for action on open science),[107] yet some see the push and pull for data access as a response to share biomedical information, and also a priority for the humanities, such as DARIAH[108] in the EU and some other in humanities,[109] and some federal agencies[110,111] in the US.

The importance of making data archives available has been voiced by many leading publishers, reflected in their initiatives that can foster effective data sharing and bring benefits to scholarly research publishing landscape.[112] Scientific Data's data policies[113] provide information on the challenging data types (e.g., experimental data, etc.)[114] data deposition (sensitive data that should be archived), guidance for selecting a suitable repository for the data including criteria for repository, and dataset updates, besides data citation. Noteworthy to mention here is OECD GSF[115] project on sustainable business models for data repositories for long-term data stewardship.

By and large, to have the relatable data available in the trusted subject-specific, generic, or institutional repositories is the general aspiration from the publishers facilitating and also supporting data policy. Typical examples include Springer

---

[101]http://ec.europa.eu/research/participants/data/ref/h2020/grants_manual/hi/oa_pilot/h2020-hi-oa-pilot-guide_en.pdf.

[102]Galsworthy, M.J. & McKee, M. (2013). Europe's "Horizon 2020" science funding programme: How is it shaping up? Journal of Health Services Research and Policy. https://doi.org/10.1177/1355819613476017.

[103]https://wellcome.ac.uk/funding/managing-grant/policy-data-management-and-sharing.

[104]https://grants.nih.gov/policy/sharing.htm.

[105]http://www.rcuk.ac.uk/research/datapolicy/.

[106]http://data.consilium.europa.eu/doc/document/ST-8791-2016-INIT/en/pdf.

[107]https://english.eu2016.nl/documents/reports/2016/04/04/amsterdam-call-for-action-on-open-science.

[108]http://www.dariah.eu/.

[109]http://guides.library.ucla.edu/data-management-humanities.

[110]http://guides.library.ucla.edu/c.php?g=180580&p=1189056.

[111]http://guides.library.ucla.edu/c.php?g=180580&p=1188862.

[112]Lin J, Strasser C (2014) Recommendations for the Role of Publishers in Access to Data. PLoS Biol 12(10): e1001975. https://doi.org/10.1371/journal.pbio.1001975.

[113]https://www.nature.com/sdata/policies/data-policies.

[114]https://www.nature.com/sdata/policies/editorial-and-publishing-policies#code-avail.

[115]http://www.codata.org/working-groups/oecd-gsf-sustainable-business-models.

Nature Research Data Policies[116] that apply to over 600 journals. Other publishers have also launched journals specifically to host data, such as GigaScience.[117,118]

Many publishers have also integrated their submission workflow with digital data repositories, such as Dryad[119] facilitating integration submission free service that allows journal publishers to coordinate the submission of manuscripts with submission of data to Dryad, and Figshare[120] allows publishers to visualize and host large amounts of data. Besides, it enables authors to share their data more easily on one hand and receive credits for their intellectual output through minting a data-specific persistent identifier, for instance DOI provided by the global DOI providers such as DataCite[121] or URI such as the ones provided by the MIRIAM Registry[122] that facilitates generation of unique and perennial identifiers. Further integration with ORCID[123] allows credit to be attributed to specific authors by connecting the authors' ORCID identifier and the dataset in question. Besides, to foster discoverability and give due credit, the publishers such as PLOS[124] and Wiley[125] now require an ORCID for all authors at the time of submitting manuscript. Also funders are riding the ORCID wave, such as Wellcome Trust, Swedish Research Council, and some others for accrediting.

8   Data Repositories

The impetus given to data archiving has led to data repositories. It is also evident that data policies, standards, principles, and procedures mandate and foster release of datasets by the authors that accompany their research manuscripts to be archived in the data repositories best suited to the specific dataset so as to ensure long-term data preservation. While selecting a repository for deposition of data in the data archive, the repository could be generalist/generalist integrated data repository,[126,127] or subject or data-type specific repository (e.g.; ArrayExpress, GEO, GenBank, EMBL

---

[116]http://www.springernature.com/gp/group/data-policy/.

[117]https://academic.oup.com/gigascience/pages/About.

[118]https://academic.oup.com/gigascience/pages/editorial_policies_and_reporting_standards#Availability%20of%20Data%20and%20Materials.

[119]http://datadryad.org/pages/submissionIntegration.

[120]https://figshare.com/services/publishers.

[121]https://www.datacite.org/.

[122]http://www.ebi.ac.uk/miriam/main/.

[123]https://orcid.org/blog/2015/10/26/auto-update-has-arrived-orcid-records-move-next-level.

[124]https://www.plos.org/orcid.

[125]http://www.wiley.com/WileyCDA/PressRelease/pressReleaseId-129824.html.

[126]https://www.nature.com/sdata/policies/repositories.

[127]https://www.nature.com/sdata/policies/repositories#general.

or DDBJ),[128,129] GODAN—Global Open Data for Agriculture and Nutrition[130]—focusing on the data ecosystem in the said area. For instance, ArrayExpress is an archive of functional genomics data that stores data from functional genomics experiments and provides these data for re-use to the researchers.

Similarly PLOS has identified and recommended a set of recognized and trusted data repositories.[131,132] An array of data repositories available at the Registry of Research Data Repositories (Re3Data)[133] and BioSharing[134] across subject areas facilitating researchers identify the repositories most suitable for their needs in terms of licensing, certificates and standards, policy, etc. Generally authors are encouraged to select repositories that meet accepted criteria as trustworthy digital repositories. In India, the domain-specific repository containing important socioeconomic data[135] is an important initiative in this direction. This data service facilitates data sharing, preservation, accessibility, and re-use of social science research data collected from entire social science community in India and abroad.

The open data initiatives at the national/international level portray that while the research data is a pre-requisite for the health of research in the data-centric world, the government data is important for socioeconomic aspects. Thus, it is equally important that the government data across sectors and departments is available for public consumption in general, to factor citizen centricity and to the researchers in particular to address developmental issues holistically. A country-wise analysis by re3data[136] lists an assortment of repositories run by institutions in various countries world over. For instance, it lists 30[137] and 844[138] repositories for India and USA, respectively. It is quite encouraging that various nations of the world, such as USA, France, UK, Denmark, Sweden, Estonia, India, and others[139,140,141,142,143,144,145,146] have created

---

[128]http://www.ebi.ac.uk/arrayexpress/.

[129]https://www.ebi.ac.uk/arrayexpress/help/GEO_data.html.

[130]http://www.godan.info/resources/research.

[131]http://journals.plos.org/plosone/s/data-availability#loc-recommended-repositories.

[132]https://biosharing.org/recommendation/PLOS.

[133]http://www.re3data.org/.

[134]https://biosharing.org/.

[135]http://www.icssrdataservice.in/#.

[136]http://www.re3data.org/browse/by-country/.

[137]http://www.re3data.org/search?query=&countries[]=IND.

[138]http://www.re3data.org/search?query=&countries[]=USA.

[139]https://www.data.gov.

[140]https://data.gov.uk/.

[141]https://europa.eu/european-union/documents-publications/open-data_en.

[142]www.opendata.dk/.

[143]http://öppnadata.se.

[144]http://Data.gouv.fr.

[145]http://opendata.ee/.

[146]https://data.gov.in/.

government data portals. Taking the case of government data portal of India[147] promoting digital initiative of India currently hosts over 140,000 datasets/resources from about 107 departments and is continuously being populated.

9 Data Citation and Scholarship

Citing publications have been an activity in perpetuity in the pursuit of science and research landscape for acknowledging and recognizing the work of other authors. The recent data phenomenon also demands to give credit where it is due. As stated in[148] "if publications are the stars and planets of the scientific universe, data are the 'dark matter'—influential but largely unobserved in our mapping process". The importance of role of citations in measuring scientific productivity and the current rewarding system in academics and research community cannot be more emphasized.[149] To be able to cite the data/dataset itself in a manner similar to citation of articles is need of the hour. Not only does data citation foster giving credit and attribution, but it also reinforces discoverability. As Borgman[150] states that "Data are representations of observations, objects, or other entities used as evidence of phenomena for the purposes of research or scholarship".

The joint declaration of the data citation principles[151] reflects the ethos of considering the data as an legitimate citable product of research and an integral part of the scholarly ecosystem supporting data re-use. Realizing the need for the use and re-use of data for re-purposing, the CODATA Task Group on Data Citation Standards and Practices[152] was formed to look into various issues of the adaptation of data citation practices. The task group published two major reports for data attribution and citation practices[153] and current status of data citation practice,

---

[147]Ibid.

[148]CODATA-ICSTI Task Group on Data Citation Standards and Practices, 2013, p. 54 Out of Cite Out Of Mind: The Current State of Practice, Policy, and Technology for the Citation of Data. Available at: https://web.archive.org/web/20160615223009/http://datascience.codata.org:80/articles/abstract/10.2481/dsj.OSOM13-043. Also published in Data Science Journal—Vol.12 (2013) pp. 1–75. https://web.archive.org/web/20160528030712/https://doi.org/10.2481/dsj.OSOM 13-043.

[149]https://www.datacite.org/cite-your-data.html.

[150]Borgman, C.L. (2015). *Big Data, Little Data, No Data: Scholarship in the Networked World.* MIT Press, 416 pages.

[151]https://www.force11.org/group/joint-declaration-data-citation-principles-final.

[152]http://www.codata.org/task-groups/data-citation-standards-and-practices.

[153]For Attribution: Developing Data Attribution and Citation Practices and Standards (2012). https://web.archive.org/web/20151102020148/http://www.nap.edu:80/catalog/13564/for-attribution-developing-data-attribution-and-citation-practices-and-standards.

policy, and technology.[154] These reports laid out the landscape of research data attribution and citation issues, practices, and policies. The crying need emerges for a framework of international agreements, practices or standards codified, national policies and practices for funding and incentivizing research. In order to promote the implementation of the data citation principles in the research policy and funding communities, the Task Group held a series of national and regional workshops[155] primarily dedicated to this focused objective with little variation in the modus operandi and has prepared a synthesis paper that integrates the findings. The workshop in India[156] focused on data science, technology, research and applications in the Indian context keeping the data citation attribution and standards protocols and practices in the backdrop. The Indian workshop in its country report highlighted three sets of recommendations for data science, data citation protocols, and stakeholders of data citation practices.[157,158] The key actions and interventions identified at the workshop were the need for brainstorming session for experts and practitioners to be held to seek and spell out the action plan and modalities for standards and citation practices protocol for various subject areas and fields as well as selection of scientific techniques that can be used in governance domain. The technical issues regarding discipline-wise specific big data and citation attributions particularly in the large datasets of science and technology need to be taken into account to facilitate implementation of policies. The need for formulation of a national-level policy in India to standardize data citation practices, DOIs, and other persistent identifiers to data for the advancement of S&T and big data science in the country was strongly advocated. This step in their opinion would make India emerge wiser on many issues related to big data. Thus, the general consensus and the agreements reached during the workshop in India regarding data citation (DC) practices primarily concentrated on four issues pertaining to (i) deeming data

---

[154]Out of Cite Out Of Mind: The Current State of Practice, Policy, and Technology for the Citation of Data (2013). https://web.archive.org/web/20160528030712/https://doi.org/10.2481/dsj.OSOM 13-043.

[155]http://www.codata.org/task-groups/data-citation-standards-and-practices/international-series-of-data-citation-workshops.

[156]http://www.codata.org/task-groups/data-citation-standards-and-practices/international-series-of-data-citation-workshops/india-data-citation-workshop-2015.

[157]Munshi, Usha Mujoo, CODATA Data Citation Workshop "Big and Open Data-Evolving Data science Standards and Citation Attribution Practices" November 5–6, 2015—Country Report—India. Un-published Report.

[158]http://www.codata.org/task-groups/data-citation-standards-and-practices/international-series-of-data-citation-workshops/india-data-citation-workshop-2015.

as legitimate citable product, (ii) DC to facilitate giving credit to where it is due, while including a machine readable identification system, (iii) flexibility of DC methods to accommodate variant practices without compromising on interoperability of citation practices, and finally (iv) DC to facilitate access to data (itself) and to associated metadata.

In order to propel more energy to the data citation standards, practices, and protocols, the Research Data Alliance (RDA) set up a Working Group (WG) for Data Citation (DC).[159] The RDA-WGDC has submitted its expected outputs[160] in the form of recommendations[161,162] regarding Scalable Dynamic Data Citation Methodology. Currently the WG is acting as a maintenance group,[163] while helping, collecting feedback and practices from adopters. The requisite feedback from adopters would facilitate in improving the recommendations, if warranted. The support for know-how and show-how of implementing the RDA Data Citation recommendations for long tail research data and CSV files is also available.[164] The Future of Research Communication and e-Scholarship (FORCE11) is organizing a Data Citation Implementation Pilot (DCIP).[165]

In publications of the journals, some of the leading publishers ask researchers to cite data descriptors using traditional literature references and to additionally cite any datasets used, where the journal supports data citations. For instance, Nature Publishing Group[166] endorses the Joint Declaration of Data Citation Principles,[167] and their data citation format is designed to adhere to these principles. The requisite procedure for providing data citations (to external main or actual datasets or other datasets) in the manuscripts is being facilitated to the prospective authors.[168]

---

[159]https://www.rd-alliance.org/groups/data-citation-wg.html.

[160]https://www.rd-alliance.org/group/data-citation-wg/outcomes/data-citation-recommendation.html.

[161]https://rd-alliance.org/system/files/documents/RDA-DC-Recommendations_151020.pdf.

[162]https://www.rd-alliance.org/rda-wgdc-recommendations-extended-description-tcdl-draft.html. (slightly more extensive draft (version) report).

[163]https://www.rd-alliance.org/system/files/DataCitation_maintenance.pdf.

[164]https://www.rd-alliance.org/group/data-citation-wg/webconference/webconference-data-citation-wg.html.

[165]https://www.force11.org/group/dcip.

[166]http://blogs.nature.com/scientificdata/2014/03/24/endorsing-the-joint-declaration-of-data-citation-principles/.

[167]https://www.force11.org/datacitation.

[168]https://www.nature.com/sdata/publish/submission-guidelines.

Similarly, GigaScience[169] also endorses and implements data citation principles[170] and software principles.[171] Even the citation of preprints wherever appropriate in the reference list is encouraged, though not taken note of in the submitted manuscript for publication for evaluation purposes. There are umpteen number of such examples from scholarly research landscape. A typical example of how a data citation looks like is given below.

Example I

The dataset[172]:

Storz, D et al. (2009):

Planktic foraminiferal flux and faunal composition of sediment trap L1_K276 in the northeastern Atlantic.

https://doi.org/10.1594/PANGAEA.724325

Is supplement to the article[173]:

Storz, David; Schulz, Hartmut; Waniek, Joanna J; Schulz-Bull, Detlef; Kucera, Michal (2009): *Seasonal and interannual variability of the planktic foraminiferal flux in the vicinity of the Azores Current.*

Deep-Sea Research Part I-Oceanographic Research Papers, 56(1), 107–124, https://doi.org/10.1016/j.dsr.2008.08.009

Example II

Another example of typical citation looks like this:

Met Office (2006): UK Daily Temperature Data, Part of the Met Office Integrated Data Archive System (MIDAS). NCAS British Atmospheric Data Centre, *date of citation.* http://catalogue.ceda.ac.uk/uuid/1bb479d3b1e38c339adb 9c82c15579d8.[174]

Typical Citation to use for data with a DOI is:

Science and Technology Facilities Council (STFC), Chilbolton Facility for Atmospheric and Radio Research [S. A. Callaghan, J. Waight, C. J. Walden, J. Agnew and S. Ventouras]. GBS 20.7 GHz slant path radio propagation measurements, Sparsholt site [Internet]. NCAS British Atmospheric Data Centre, 2003–2005, 1st April 2011, https://doi.org/10.5285/E8F43A51-0198-4323-A926-FE6922 5D57DD.[175]

---

[169]https://academic.oup.com/gigascience/pages/editorial_policies_and_reporting_standards# ReportingStandards.

[170]https://www.force11.org/group/joint-declaration-data-citation-principles-final.

[171]Software Citation Principles https://peerj.com/articles/cs-86/.

[172]https://doi.pangaea.de/10.1594/PANGAEA.724325.

[173]Ibid.

[174]http://help.ceda.ac.uk/article/102-data-citation.

[175]Ibid.

In order to improve the access to data on the Internet, several organizations signed a Memorandum of Understanding (MoU) to facilitate registration of research datasets and assign unique identifier to discover the data/datasets.[176] This endeavor resulted in the establishment of a nonprofit organization came to be known as DataCite[177] that enables organizations to register research datasets and assign persistent identifiers to them. Since its establishment, a number of research datasets have been registered on DataCite and are being populated with more and more datasets on continuous basis. One can easily find out the numbers from the summary statistics that includes data from all the data-centers that use DataCite DOIs and is available on its DataCite statistics site.[178]

10 Data Rescue for Data Re-Use

The research landscape across subject domains possesses a rich heritage of data that encompasses both in the electronic as well as analogue format. The argument that if the analogue datasets are not archived properly they are susceptible to permanent loss. The accessibility to research data across domains without barriers has added a new dimension to the importance of data management for its subsequent use. The recent focus and initiatives on keeping the data safe, for instance, rescue data that is at risk for re-using/re-purposing to generate new knowledge is at its best shove. Just like in the text archiving, we migrated from print to digital and then took collab-orative initiatives[179,180,181] to preserve access to content and archive for future generations in a trusted digital environment, lest we are left with unreadable media floating around our places.

---

[176]Brase, Jan. DataCite—A global registration agency for research data. Available at: https://www.ratswd.de/download/RatSWD_WP_2010/RatSWD_WP_149.pdf.

[177]https://www.datacite.org/.

[178]https://stats.datacite.org/.

[179]Lots Of Copies Keep Stuff Safe https://www.lockss.org/.

[180]https://www.clockss.org/clockss/Home.

[181]http://www.portico.org/digital-preservation/.

A number of[182,183,184,185,186,187,188,189,190,191,192,193,194,195,196] reports and empirical studies and surveys have focused on the issue of data recovery and access and reported on numerous "data rescue," "data refuge," and "guerrilla archiving" events that have taken place in various countries of the world during the recent

---

[182]Varinsky, Dana. (2017). Scientists across the US are scrambling to save government research in 'Data Rescue' events. Business Insider, Feb. 11, 2017. http://www.businessinsider.com/data-rescue-government-data-preservation-efforts-2017-2.

[183]Science News Staff. (2017). A grim budget day for U.S. science: analysis and reaction to Trump's plan. Science, Mar. 16, 2017. https://doi.org/10.1126/science.aal0923.

[184]Temple, James. Climate data preservation efforts mount as Trump takes office. MIT Technology Review, Jan. 20, 2017. https://www.technologyreview.com/s/603402/climate-data-preservation-efforts-mount-as-trump-takes-office/.

[185]Khan, Amina. Fearing climate change databases may be threatened in Trump era, UCLA scientists work to protect them. Los Angeles Times, Jan. 21, 2017. http://www.latimes.com/science/sciencenow/la-sci-sn-climate-change-data-20170121-story.html.

[186]Harmon, Amy. Activists rush to save government science data—If they can find it. New York Times, March 6, 2017. https://www.nytimes.com/2017/03/06/science/donald-trump-data-rescue-science.html.

[187]Yarmey, K. and Yarmey, L. (2013). All in the Family: A Dinner Table Conversation about Libraries, Archives, Data, and Science. Archive Journal, Issue 3. http://www.archivejournal.net/issue/3/archives-remixed/all-in-the-family-a-dinner-table-conversation-about-libraries-archives-data-and-science/.

[188]Herrmann, Victoria. (2017). I am an Arctic researcher. Donald Trump is deleting my citations. The Guardian, Mar. 28, 2017. https://www.theguardian.com/commentisfree/2017/mar/28/arctic-researcher-donald-trump-deleting-my-citations.

[189]Anderson, William L., Faundeen, John L., Greenberg, Jane, & Taylor, Fraser. (2011). Metadata for data rescue and data at risk. In Conference on Ensuring Long-Term Preservation in Adding Value to Scientific and Technical Data. http://hdl.handle.net/2152/20056.

[190]Downs, Robert R. & Chen, Robert S. (2017). Curation of scientific data at risk of loss: Data rescue and dissemination. In Johnston, Lisa (Ed). Curating Research Data. Volume One, Practical Strategies for Your Digital Repository. Association of College & Research Libraries. https://doi.org/10.7916/D8W09BMQ.

[191]Griffin, R.E. (2015). When are old data new data? GeoResJ, 6: 92–97. https://doi.org/10.1016/j.grj.2015.02.004.

[192]Thompson, C.A., Robertson, W. D., & Greenberg, J. (2014). Where have all the scientific data gone? LIS perspective on the data-at-risk predicament. College & Research Libraries, 75(6), 842–861. https://doi.org/10.5860/crl.75.6.842.

[193]Molteni, Megan. Diehard coders just rescued NASA's Earth science data. Wired, Feb. 13, 2017. https://www.wired.com/2017/02/diehard-coders-just-saved-nasas-earth-science-data/.

[194]Molteni, Megan. Old-guard archivists keep federal data safer than you think. Wired, Feb. 19, 2017. https://www.wired.com/2017/02/army-old-guard-archivers-federal-data-safer-think/.

[195]Douglass, K., Allard, S., Tenopir, C., Wu, L., Frame, M. (2013). Managing scientific data as public assets: Data sharing practices and policies among full-time government employees. Journal of the Association for Information Science and Technology, 65(2): 251–262. https://doi.org/10.1002/asi.22988.

[196]Tenopir, C., et al. (2015). Changes in data sharing and data reuse practices and perceptions among scientists worldwide. PLoS One, 10(8): e0134826. https://doi.org/10.1371/journal.pone.0134826.

times. The lifecycle stages of data recovery, from location of assets to new science, are data rescue. Just like text archives, data is at risk, if data is not available/accessible/retrievable in digital format. Still, in many cases data is available in individual silos thus practically forgotten data (perhaps only author knows it exists), hence not accessible for use/re-use. Irrespective of the relentless parading sophisticated technologies available for archiving/curating, many important datasets are discrete. This reflects that our attitudes are overruling our ability. For instance[197] loss of priceless heritage of the National Museum of Natural History, India due to the fire in 2016 is a glaring example of the sort.

Globally a number of initiatives led by different stakeholders are on track for data asylum to discover and recover research data at risk of disappearing. For instance, many governments have created huge datasets and floated data on government portals and many departments self report their raw data. Typical examples could be www.data.gov.in or data.gov (USA) and the like. However, there are hundreds of government FTP servers and thousands of government Websites which may contain links to data, but only a few of these FTP/HTTP resources have machine-readable metadata. What is important is to have much better systems in place so that the metadata (machine-readable) can be generated for discoverability.

To give impetus and drive the energy in re-creating the analogue datasets that can critically contribute to the studies for unraveling thus veiled trends due to non-accessibility of data, some concerted efforts have cropped up. Some of the legacy data are also floating around the places in unknown/unrecognized formats due to technological obsolescence and also lack metadata, hence left out unused. The typical examples of such efforts include constituting CODATA Task Group on Data At Risk (DAR-TG).[198] DAR-TG's core objective was to raise awareness of the high scientific potential of all heritage data through various well-defined activities including various conference sessions since 2015. The imperative nature of data rescuing efforts has witnessed attention from across the corners. The CODATA's DAR-TG became affiliated in 2015 to the Research Data Alliance in the form of an Interest Group christened as RDA Interest Group for "Data Rescue".[199] The Data Rescue involves two strands of data management[200]—(i) the recovery and digitization of analogue data (not born digital), and (ii) adding essential value to archives of (born digital) electronic ones (e.g., metadata, format information, access).

The setting up of the groups investigating Data At Risk and delivering models for plausible recommendations and actions are the core aspirations behind setting up of such Working/Interest Groups. The RDA Interest Group on Data Rescue

---

[197]Munshi, Usha Mujoo. Past Forward: Can we relive the (lost) treasure. Presentation made during CODATA-RDA Workshop on Rescue of Data at Risk, during 8–9 September, 2016 NCAR, Boulder, CO, USA.

[198]http://www.codata.org/task-groups/data-at-risk.

[199]https://www.rd-alliance.org/groups/data-rescue.html.

[200]Ibid.

(RDA-IG on Data Rescue) has come out with Guidelines for Data Rescue.[201] To boost both data rescue and data at risk activities, the RDA-IG has appointed two data share fellows, one each for Data Rescue and Data at Risk, respectively.

Even the identification of what data is at risk and needs to be rescued is an important activity to be taken up in project mode. To this end, already the Data-At-Risk Initiative (DARI) a project (2010–2013) of the Committee on Data for Science and Technology (CODATA) Data at Risk Task Group (DAR-TG) attempted to create an inventory of valuable scientific data that are at risk of being lost to posterity.[202] At individual level, some domain specific guidelines for data rescue have also cropped up.[203]

The collaborative endeavors have been the key driver for pushing new and innovative ideas that boosts efficiency and brings about transparency, while creating a robust system for sharing best practices. It is thus natural that collaborative linkages are necessitated for identifying, rescuing, and preserving data at risk. The Earth Science Information Partners (ESIP)[204]—an open, networked community, connecting science, data and users has come out with a set of recommendations for identifying and preserving data at risk.[205]

11   Data Infrastructure for Data Discovery, Sharing and Re-Use

Big data is likely to pave way for data to live permanently on archival media like its contemporary print and paper archives. Data access and sharing depend on knowledge infrastructure. The development of data infrastructure to create conducive and innovative data environment for efficient flow of data is needed to ensure broader benefits to the economy and society at large. The data infrastructure would thus entail whole set of entities that includes people, processes, and technological infrastructure as an enabler together with the skill set to buttress an array of research endeavors and its application to promote and sustain fineness of research landscape and fulfill Third Mission. In order to make that happen, the strategic threads have to be woven to integrate discrete elements and create a cohesive structure or system by collating and networking people, processes, and technology to ensure return on past investment. Also it has to be worked out strategically about how to motivate and incentivize the researchers for data sharing,

---

[201]https://www.rd-alliance.org/system/files/documents/Guidelines-2_0.pdf.

[202]http://cci.drexel.edu/mrc/projects/dari/.

[203]http://www.wmo.int/pages/prog/hwrp/publications/guidelines_hydrological_....

[204]http://www.esipfed.org/.

[205]Matthew S. Mayernik, Robert R. Downs, Ruth Duerr, Sophie Hou, Natalie Meyers, Nancy Ritchey, Andrea. Stronger together: the case for cross-sector collaboration in identifying and preserving at-risk data. Available at: http://www.esipfed.org/press-releases/stronger-together.

scaling up the procedures and processes for interoperability, curation and collaboration across segments, be it Government, academia, R&D in public/private sector, or the like. It is also important to make use of the existing computational infrastructure by creating simple yet effective interfaces that permits compliance with sophisticated scalable technical developments. The e-infrastructure thus is expected to take care of the critical issues, such as storage (voluminous data), tools, and techniques (including software/s) that facilitate complex computational strategies and ensuring secured access. The infrastructure has to be robust such that in immediate future the data becomes an integral and occupies a decisive position within the scholarly communication environment, the same way as its textual research resources have robustly settled in, to have a full circle DMP.

Many nations have developed infrastructure to support data discovery, access, and usage in a trusted environment. In India, similar initiatives are underway to facilitate infrastructure so as to support data access and sharing. There are several success stories of use of big data in Governance in India.[206] The National Informatics Centre (NIC) has developed data portal to share data under National Sharing and Accessibility Policy, though the need for developing algorithms, getting real-time data, and quick analysis remains to be addressed. The various historical/live datasets available on data portal of India and various visualizations are being made out of those datasets.

The Government of India has already initiated a process of putting open data in public domain, and the open data platform is hosted at http://data.gov.in/. Already around 107 government agencies and departments in India are sharing their data on the platform. The Government departments need to publish their shareable data proactively on this platform. Sometimes it requires to contact the specific government department to release datasets. The issues pertaining to social science data generated by Government agencies have been taken up by the Indian Council of Social Science Research (ICSSR) with concerned organization, i.e., Ministry of Statistics and Programme Implementation. Furthermore, ICSSR Data Repository-Policy Framework exists, and the procedures to cite data from ICSSR Data Service are clearly specified. The ICSSR Data Repository-Policy Framework which is available online includes data preservation, data access, metadata, data submission, and content coverage. In addition, there are State Data Centres (SDCs) and National Data Centres (NDCs) that are providing e-district applications issuing socioeconomic data.

At national level no persistent identifier system is available to all researchers in the country. It is though important to get hold of a persistent identifier, by a centralized agency like NIC or some other agency. It is thus imperative that prioritizing data for discovery without duplicating the efforts by making use of DOIs is a core requirement. Also, metadata needs to be just and absolute for promoting data

[206]Hickok, Elonnai; Chattapadhyay, Sumandro, Abraham, Sunil (Editors). Big Data in Governance in India: Case Studies. The Centre for Internet and Society, India. Available at: http://cis-india.org/internet-governance/files/big-data-compilation.pdf.

discoverability, access, and use. This will help in furtherance of populating data in the data portals in a trusted environment that will greatly enhance accessibility and discoverability of datasets.

12 Data Science Paradigm—International Initiatives

The systemic analysis of the data warehouses not just gives insights but helps in deep learning and strategic decision-making. At the same time it instigates the development of advanced data-driven capabilities, and collectively they contribute toward value addition to the business endeavor. Many international initiatives to foster data governance plans are maturing to realize their vision and mission. For instance the Global Pulse initiatives[207,208] to support sustainable development goals is one such example. The endeavors of ICSU's World Data System (WDS)[209] and CODATA[210] to enable equitable access to trusted data itself, data resources, services, and products globally are other examples. World Economic Forum (WEF)[211] on various facets of harvesting the big data insinuates value extraction from big data and foreseeable data driven economy. IDC's[212] prediction about growth in applications for predictive analytics and technology adoption for continuous analyzing streams of events also connotes both pull and push for Big Data in big way across sectors. As reported by Ashish Nadkarni and Dan Vesset in IDC's big data technology and services forecast report[213] "Revenue for big data services, which consists of professional and support services, is estimated to grow at a CAGR of 23.9% from 2015 to 2020 and reach $15.2 billion in 2020". The global call for big and open data has resulted in action-oriented initiatives. Open data initiatives have resulted in making over 49 nations (with 2074 repositories),[214] including India to have established open data portals to factor citizen centricity. This has resulted in development of making of open source software for big data analytics that has since taken off well.

---

[207]http://www.unglobalpulse.org/.

[208]http://unglobalpulse.org/sites/default/files/IntegratingBigData_intoMEDP_web_UNGP.pdf.

[209]https://www.icsu-wds.org/.

[210]http://www.codata.org/.

[211]https://www.weforum.org/agenda/archive/big-data/.

[212]https://www.idc.com/getdoc.jsp?containerId=IDC_P23177.

[213]Ashish Nadkarni and Dan Vesset. Worldwide Big Data Technology and Services Forecast, 2016–2020. IDC, Dec 2016, p.9. Doc# US40803116 Available at: https://www.idc.com/getdoc.jsp?containerId=US40803116.

[214]http://www.re3data.org/browse/by-country/re3data.org.

13   Case for India

Like any other developed or innovatively developing country, India is making headway toward big data applications in both research landscape and addressing socioeconomic issues for sustainable development. So far as the entrepreneurship development is concerned, as per the joint study by NASSCOM and market intelligence firm Blueocean, the size and reach of the Indian analytics market are expected to double from USD 163 million in 2014 to USD 375 Million by 2018.[215] Expected analytics market in India, which clocked revenues of about USD 1 billion in 2013–14, is expected to be more than double it to USD 2.3 billion by the end of 2017-18 fiscal. Over 500 companies are operating in this segment in India. The total analytics employees are over 29,000, while the domestic market focused on 5,000 in 2014, whereas the market leader US is expected to have a shortage of 140,000–190,000 analytics professionals by 2018, which opens up a huge opportunity for product and service companies in India.[216] The scenario divulges that industrialization of analytics in Indian enterprises is vital to drive long-term value and be the market leader in analytics, alongside the fastest growing economy, while retaining that status in the years to come.

Of late, quite noteworthy initiatives of digitalization for inclusive development have surfaced to the top. We will take a look at a few of them here, though our main focus would limit here to data intensive efforts for re-use and re-purposing of the data. The typical examples include flagship program of Government of India "Digital India" program[217] to digitally empower the society and boost knowledge economy holistically. One of the massive social welfare schemes in the digital initiative of Government of India focused toward the poor and most vulnerable sections of society is Aadhaar.[218] It offers a unique opportunity to the government to streamline their welfare delivery mechanism and thereby ensuring transparency and good governance.

Department of Science and Technology (DST) Government of India initiative to facilitate promotion and development of data science, technology and applications and related ecosystem in the country under its "Data Science, Technology, Research and Applications (dASTRA)"[219] program is one of the examples of promoting data science ecosystem in the country.

There is a need for lively data science ecosystem for India. In this data ecosystem, the integration of people, processes, and technology is a pre-requisite. For instance Government and user agencies, academic institutions, industry and

---

[215]NASSCOM Report on Institutionalization of Analytics in India: Big Opportunity, Big Outcome, 2014. http://www.nasscom.in/nasscom-big-data-summit-redefining-analytics-landscape-india.

[216]https://www.slideshare.net/blueoceanmi/industrialisation-of-analytics-in-india.

[217]http://www.digitalindia.gov.in/.

[218]https://uidai.gov.in/your-aadhaar/about-aadhaar/.

[219]https://www.cdc.org.in/UserFiles/File/2015/Project/Big_data/DRAFT%20REPORT%20V3.pdf.

startups, collaborations/inter-disciplinary approach, systems perspective and societal applications all have to gel well. A few initiatives in the academia and R&D
sector regarding data science are happening. For instance, institutions such as
Indian Institute of Science (IISc), Bangalore, Indian Statistical Institute (ISI),
Kolkata, and Indian Institute of Technology (Bombay, Madras, Delhi, Kanpur,
Kharagpur) are engaged in the R&D of various segments of data science, e.g.,
machine learning, algorithms, statistics, sampling, linear algebra, convex optimization, game theory, signal processing, large visualization, cloud and distributed
computing, and the like. Initiatives from various segments that result in voluminous
datasets and analytics such as AADHAR (biggest highway for facilitating delivery
of citizen centric services meticulously),[220] Flipkart,[221] Strand Life Sciences
(Algorithms for life),[222] and many more are being nurtured, and the segment is in
continuous flux of evolving and so are the processes and the procedures.

14 Conclusion

Data is a strong pillar for sustainable development. Building data-centric applications while augmenting infrastructure to have a robust system of data management
(from generation to organization to dissemination, discovery and access) for data
re-use and re-purposing is inevitable. In the present complex systems, it is intrin-
sically difficult to model economic complexity, country's industrial composition, or
socio-political behavior. Thus, it is apparent in the current scenario that
intuition-based policy making is not going to help any more, but the smart and
sustainable decisions require evidence-based policy formulation that factors data
centricity. To make use of big data analytics for public good, developing infrastructure to cater to the data tsunami is extremely of great magnitude. This upsurge is
anticipated mainly in the application layer of the big data needs.

The focus on conceptual, logical, and physical data models that support research
and build robust systems of governance for societal good is the need of the hour. At
the same time, the role of research funding and policy community in implementing
data policies and practices is extremely important, and the enactment of the
National Data Sharing and Accessibility Policy (NDSAP) of Government of India
demands its taking appropriate steps.

The discussion of several events has culminated in focusing on some common
tangible outcomes regarding international linkages and collaboration, human
resource development, mandatory deposition of research data by public funding
bodies, inter-disciplinary research, standards and protocols for data research management using a holistic approach, for making big data science landscape really big.

---

[220]https://uidai.gov.in/your-aadhaar/about-aadhaar.html.

[221]http://flipkart.com.

[222]http://strandls.com/bioinformatics/.

For instance in one of the events of big data,[223] the emerged recommendations from the detailed deliberations were directed towards the following: (i) the collaboration between S and T sector with the social science sector should be further strengthened and such an activity should be in PSA mode; (ii) secondly, international collaboration should be emphasized and more strengthened so that progress can be more fast and also more or less at the international level. South-south collaboration is very important; (iii) the issue that needs to come out very strongly is about the scientific data which emerges out of projects sponsored by Government agencies like DST and DBT and other Government funding bodies. All data Principal Investigators (PIs) must be asked to put the data in public domain so that other users can use it. In this, of course, U.S. and Europe are in the forefront wherever public money is used. For India, it is high time to make sure now that the infrastructure for hosting such data is available and can be easily managed; (iv) the other important critical issue which emerged, is a dire need to prepare manpower which is specialized and has capability and expertise in the field of data science and technology. Therefore, some universities/colleges and other such organizations should be encouraged to initiate, if not a primary degree but may be M. Tech level or short-term courses in data science which will go a long way in bringing the subject up. A unanimous perspective and generally agreed norm is that the big and open data concept should he initiated from the initial learning level, and big data should come to classroom for better teaching. Also, now in government sector we need to set up data systems and make their accessibility available through devices such as mobile phones at grassroot level.

While we talk about data in the research arena, the data itself should be considered a legitimate citable product of research. By doing so the way we were wary of giving credit to people where it is due and protecting their Intellectual Property Rights (IPR), the same way we need to evolve robust system for data citations. Data citations should facilitate giving scholarly credit and legal attribution to all contributors of the data. Also, data citations should facilitate access to the data themselves and to the associated metadata. A data citation should include a method for identification which is machine-actionable (for instance, use of DOI). In India, no such standard practice is currently in vogue.

The issues regarding security and incentivizing the data producers are debated to suggest feasible and doable mechanisms. The idea that data should always conform to legal, ethical, and regulatory frameworks including appropriate acknowledgments is an underlying chore. It should be obligatory to recognize through citation and acknowledgment of the original creators of the data. This would result in benefits such as providing security, protecting intellectual property, and incentivizing the data contributors by giving credit, where it is due, coupled with effective retrieval, use, and re-use of data by others. Therefore, while data

---

[223]Munshi, Usha Mujoo, CODATA Data Citation Workshop "Big and Open Data-Evolving Data science Standards and Citation Attribution Practices" November 5–6, 2015—Country Report—India. Un-published Report.

supporting publications should be accessible by the public, it should be in citable form. This ensures and aids scientific replication and validation, providing permanent and reliable information on data sources. The underlying principle being the citing data is vital for all subject domains.

The question of how to make sure we have in place a system for data citation is very important to ponder over. Data citation methods should be sufficiently flexible to accommodate variant practices but should not compromise interoperability of data citation practices. While the provision of data for citing practices is important but the fundamental issue of addressing the problem of cost for citation of data must also be taken into serious consideration.

The role of research funding and policy-making community in implementing data citation policies and practices is extremely important, and the enactment of the National Data Sharing and Access Policy (NDSAP) of Government of India demands its taking appropriate steps. The deliberations of the event[224] led to the suggestions of carrying out a small study that should clearly portray as to how a national system on this can be evolved. Out of this event, the experts opined to recommend to government bodies like DST, DeitY, or NIC and other bodies who are involved in this to set a group or task force to work on the recommendations which can eventually be accepted by everybody.

The other core issues regarding data in the big data landscape revolve around flexibility, machine readability, and metadata issues along with inter-operability factors that lie at the heart of making things happen so as to promote discoverability and sharing of data. None will argue that privacy should be maintained and cannot be disclosed for data mining in many domains such as scientific, business, healthcare etc. Besides, there is a great need to evolve techniques for privacy preservation of data so that it can be utilized for further analysis. Such concerns are being voiced at many forums.

The policy (NDSAP) principles need to be wrapped with a continuous improvement loop as insights are fed back to improve the data analysis models as well as the process. These in turn can only drive the quality and completeness of data to further press forward future analysis.

Thus, spinning trio of philanthropy of promoting sharing and access to streaming data, technology to share tools and techniques for data mining, interpretation and subsequent projections, and developing expertise by capacitating human resources in making big data phenomenon even bigger through innovation in creating a knowledge-driven research infrastructure is highly necessitated. These initiatives will foster sustainability of such interventions for sustainable development.

The world over it is evolving to bring about breakthrough revolution in analytics for drawing foolproof inferences for smart decision-making and informed feedback. The Indian market is still in early stages of adoption of analytics, and there is a need to industrialize use of analytics to derive long-term value. However, with surplus talent, established infrastructure, and a mature ecosystem, India is on its way to

---

[224]Ibid.

become a global hub for analytics. Big data will continue to mature for providing evidence-based measures to change the power structures for societal good through responsible research and innovation with robust data and analytics system in the backdrop.

Open Data Infrastructure for Research and Development

Neeta Verma, M. P. Gupta and Shubhadip Biswas

1 Introduction

Value of the research data is immense, and opening research data can foster innovation, new insights and discovery. Open research data related to health, energy, physics, mechanics, social science, economy, etc., enable these raw data to contribute in the respective domains far beyond the primary analysis. Raw research data are not only being used to validate the original analysis, but also it can be helpful to discover interrelated or newly defined hypotheses, especially when associated with other openly available data.

In recent past, it has become more common for researchers to publish their research data as open data. Funding agencies increasingly require the research data (and publications) resulting from funded research projects to be published open access. However, open data access is not yet standard practice in most disciplines, and there is no culture of data sharing and reusing among researchers. Even when researchers in these fields publishes their data in the repositories and archives, the data are usually difficult to find and to access. Many international bodies are promoting open access of publications and data, like the Open Knowledge Foundation [27], Open Data Institute [26], OpenAIRE [28], OAPEN [25], and Knowledge Exchange [18]. There is often no proper infrastructure for central repository or registry of data; unavailability of central repository of data is one of the major

N. Verma (✉)
National Informatics Centre, New Delhi, India
e-mail: neeta@nic.in; neeta@gov.in

M. P. Gupta
DMS, Indian Institute of Technology, Delhi, India
e-mail: mpgupta@dms.iitd.ac.in

S. Biswas
Open Government Data Project, Delhi, India
e-mail: shubhadip.biswas@live.com

© Springer Nature Singapore Pte Ltd. 2018
U. M. Munshi and N. Verma (eds.), *Data Science Landscape*,
Studies in Big Data 38, https://doi.org/10.1007/978-981-10-7515-5_2

33

challenges, and using technology, this need can be addressed by setting up a central platform for participation and collaboration, which will allow all the stakeholders to interact and explore data provision and thus enhance the potential for value addition and innovation. Open data infrastructure should enable researchers to publish their data for its use and reuse. Setting up such an infrastructure is a cost and resource-intensive effort. There is a need for policy, governance, and financial support for setting up such an infrastructure and maintain it over the years.

The paper delves into various aspects of open data infrastructure (ODI) that includes:

- Policy formulation of ODI
- Fund support for ODI
- Implementation of ODI
- Metadata and data standards
- Data use license
- Open data ecosystem
- Data citation mechanism

Focus should also be on the improvement of the quality of information, the formation and establishment of open data culture, and the delivery of the tools and mechanism to use data. A technological robust infrastructure is essential which helps all the stakeholders to make sense of data and to ensure community participation. Proper collaborative framework can enable open data to go beyond the current level of data access and research and development to foster innovation and socioeconomic growth.

2 Literature Review

Open data have become very important in research domain. Estermann demanded that the open data initiative in academic sphere started around fifty years ago [9] with the publication of the first scientific journal in 1965, i.e., philosophical transactions of the royal society, it had the policy of founding concepts escorted by the evidence on which it was constructed (i.e., data) [3]. Opening research data evade the effort to rework on the same model, provide substantiation that the methodology used for the research was accurate and correctly implemented, display answerability of the researcher, and generate the prospect of new research findings [11], not sighted by the previous researcher(s) [12]. Several journals, particularly science journals, are in support of opening experimental data [33] so that they can be reused, reproduced, and authenticated. Similarly, disclosing research data are nowadays a prerequisite for data management planning and release research application's policy [6]. Few journals also demand for the consent that research data would be shared on request [13], while for some other journals it is prerequisite.

In view of the political, social and economic factors, and academic significance and prospect of its broader application, the idea of open data has started drawing attention of the researcher community, which is evident through the emergent topic in published research papers and main topic of discussion during academic conferences. These events have experienced a sharp growth in recent years in research projects.

Open Government Data (OGD) have also been drawing a rising notice and concern of both researchers and activists from various branch of learning, such as information technology, management studies, social and political sciences, and law, due to its broadly recognized prospect to create public value through thrusting economic growth and innovation, and methodical research, and by promoting openness and significant evidence-based diplomatic dialogue [7, 15, 32]. The idea of open data is strongly related to inventive capacity and metamorphic power [8].

Various research and studies highlight that the citizen is generally open to the idea of Open Government Data but have unclear concepts about how it may relate to their lives. A study conducted by the Pew Center for Internet and American Life found the citizen rarely connect Open Government Data with collective ideas [19]. That is, the citizen has limited awareness of how to retrieve and interpret data. Consequently, researchers have generally aimed on "intermediaries of open government data," i.e., download, infer, and maneuver data [20, 30]. These data experts are important in our current time when government uses data for decision making that impact the citizen's life. Yet, majority is unknown about how intermediaries of Open Government Data can materialize the citizen's benefits of open data and are able to link it with uses useful for the community [11]. Technology can enable better discovery, ease of access, and innovative and wide use of open data; many countries in the world have already set up open data portal for this purpose [35, 36]. Open datasets should be released on portals through a structured workflow. Using technological infrastructure, platform can enable better discovery, ease of access; open data could be easily downloaded in open formats or consumed via application program interfaces (APIs) [37]. Open data platforms can be an important instrument to engage with citizens and communities to develop new products and services using open datasets.

3 Key Elements of Open Data Infrastructure

3.1 Policy Formulation of Open Data Infrastructure

To build any sustainable open data infrastructure, there is need for a robust policy framework. Policy provision has strong impact on strategy formulation and implementation of any open data program. The issue of research data publication and its citation has been highlighted by many publishers and journals in their style guides, predominantly from the scientific domain. American Sociological

Association has mentioned about machine readability of data and persistent identifier of references for future access; University of Chicago Press has emphasized on scientific databases; National Library of Medicine has publicly opened part of the data on the Web; the Council of Science Editors has put their complete research databases on the Web [14]; and many other research organizations like Nature Scientific Data, GigaScience, F1000Research, and Geoscience Data Journal have initiated publishing research data in open domain [31]. Although mandating of publication of research data is an efficient strategy [22], some other strategies have also been adapted by few research bodies, "acknowledge Open Practices" badge by Center for Open Science [17], which is used by the journal Psychological Science. Digital Object Identifier mechanism can be implemented to enable accurate citation, which would enable to track usage of the data. Academic credit of data through journal would also inspire researchers to proactively deposit more data in the repository. Using this method, peer review of datasets for data journal would take less efforts and would also reduce the delay of publishing research papers. Data managers and data contributors would get credit, which was not possible earlier. So, policy and guidelines would result into better discoverability and conceptualization of data.

3.2 Fund Support for Open Data Infrastructure

A major obstacle for open data sharing is sustainable funding; most of the government bodies or research organizations do not have any provision of funding for data publishing or data preservation, and once the research is over. Funding data activities must be a fundamental part of the scientific research effort. Development projects of open data infrastructure may be realized through grant funding by collaborations between research organizations and government bodies [16]. Government bodies can fund such initiatives, and there are many research agencies who can fund and be benefitted from promoting the optimal use and reuse of data in which funds were invested. They can do this by encouraging good data practices, investing in data infrastructure, and raising open data awareness.

3.3 Implementation of Open Data Infrastructure

The life cycle of research data contains all the stages of data from data collection for a study to sharing and reuse. Research data have a longer span of life than its origination, i.e., research project. At every stage of the life cycle, researchers may enhance the access of data, and other researchers may use the data in new research projects. Life cycle of research data originates with the study concept, after finalizing the concept, data collection phase starts. After collecting and collating data, data are then analyzed to get research findings. Processed data can be stored in some location (i.e., data repository, archive etc.), where it can then be discovered and accessed by other researchers in future. Share and access of research data lead to the reuse of data [1], and it makes a continual loop back to the data discovery and access stage, where the redistributed data are stored and shared for open access. The UK Data Archive provides an structured description of the data cycle which has been shown in Fig. 1.

Nowadays, data management is a crucial task of the management in any organization. Process of data management is regulating the generated information during a research project. Data management is an integral part of any research projects, and there is an increasing demand for scholars who can plan and implement standard data management practice for research organizations.

Implementation of open data infrastructure requires formulation of policies, regulations, planning, execution, and management programs to regulate, conserve, deliver, and boost the value of research data and information. This idea became popular with evolution of technology from sequential processing to random access processing. The involvement of different stakeholders across world makes the implementation of open data infrastructure more challenging, which can be addressed only through the proper policy and technical implementation.

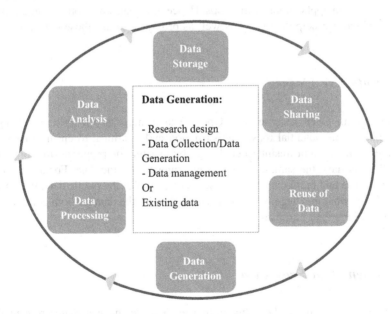

Fig. 1 Research data cycle

3.4 Metadata and Data Standards

Metadata are the key enabler to make data discoverable, utilizable, and comprehensible. To support data detection and data credentials, there are various standards, and formats of metadata have been developed over time. Metadata consist of few key elements which can be characterized according to their functionality. A standard metadata support some defined functions and describe elements to make those comprehensible [23]. Published metadata specifications with all the definitions, standards and formats should be held in a central place, and it can be published as reference file on the website or it can be kept in an accessible metadata registry.

Standardized data field is usually not followed while publishing data, as the variable names, units, and types vary across different datasets. As a result, mining of data, comparison or correlation of data is not feasible. It requires a lot of processing of data to make it ready for analysis and mashup, which is major constraint in use of data by researchers, developers, analysts, and even civil society. Hence, it is also essential that standards for metadata also be defined for open data so that the data are available in globally accepted standard format which can be used by any application confirming to those standards [34].

Better metadata or explanation of the data leads to better discovery and better reuse of data via applications or mashups. Hence, metadata are another prime aspect of data which can help data users as well as providers to use those datasets.

3.5 Data Use License

Associating an Open License with Open Data is necessary to ensure the legal grounding for its potential reuse, redistribution and is critical to ensure that such data are not misused or misinterpreted (e.g., by insisting on proper attribution), and that all users have the same and permanent right to use the data. For a data user wishing to use and build data products/services on top of the public data, they need assurance of what they legally can and cannot do with the data for both commercial and non-commercial purposes.

3.6 Open Data Ecosystem

A vibrant ecosystem can be a big influencer and can have a major role in the success of any open data initiatives. Open data can help to construct this thriving ecosystem that would create vast opportunities in research and development [5]. Open data ecosystem can be described as the provision to supply and consume open government data. Identification and involvement of the key actors would be

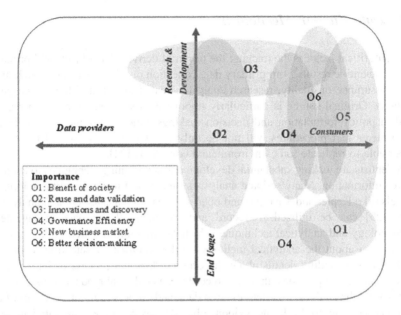

Fig. 2 Data sharing mechanism

foundation of a perfect ecosystem [20]. Policy and framework need to be developed to make the ecosystem function. A sharing mechanism has been shown in Fig. 2.

There are two sides of Open Government Data ecosystem supply side and consumption side. Data generator and providers are falling in the supply side; intermediaries and the end users will fall into consumer side.

Supply Channel. Main source of open data is governments, as they compile and generate huge data in their day-to-day operations and delivery of services. Another major source of open data is research community that may be research project or individual research. These real-world datasets require a strong data management infrastructure to supply uninterrupted data services. Therefore, data organization model would be required where data providers can share data in open format.

Consumption Channel. Consumers of open data can be divided into two major categories: One is intermediaries, and another one is end users. Intermediaries are those who will add value to the raw data shared by data providers. Application developers, researchers, journalists, data evangelists, civil society/NGOs, data scientists, and policy makers can fall into intermediate consumers of data. This group can play crucial role in creating sense of, and creating value out of, raw open data. Government, citizens, and individual users are the end users in open data ecosystem, who will use the data, products and services build out of the data. Interaction among all the players is very crucial, and understanding each stakeholder is important, as it can help to understand the potential and usage of open data [20].

3.7 Data Citation Mechanism

Another critical factor is the tracking the origin of derived data objects and primary data or scientific results from primary data collection or initial research, which may affect assurance of quality, research analysis, created model, and finally the publications. Original source is particularly important to substantiate research results used in policy formulation and decision makings, where reproducing of original experiments and procedures will not be only feasible, but also would be nearly impossible to replicate same environmental conditions [24].

Scientists are making substantial development in designing techniques to capture source information. Many scripted analysis systems built using scientific workflow models (like Kepler and Taverna) and open source data management and analytical tools [29] can be utilized to record and maintain all the data management methodology and analytical techniques which resulted into research findings. Vital information about the analytical methods can be recorded by scientific workflow applications, including details about the original data and process of its transformation, theses can provide a thorough record of an analytical model and its outputs. In this way, the analytical model, research data, and outputs turn out to be part of an information base to back the evidence-based science, to enable efficient and informed decision making in research domains [10]. New research in this domain highlights that e-Science associations are possible through the open data, and that these collaborations can be efficiently tracked to provide automatic attribution facility, i.e., where original data owner can be given credit via attributions which are resulting from provenance trace models [2, 4, 21].

4 Conclusion

The potential of open data infrastructure is enormous. Open data are a critical resource for academicians and researchers to derive a lot of insight in socio-development phenomena and in scientific research. Data are more likely to be proactively shared when the data providers and users both feel the necessity of data dissemination. There is strong sense of collaboration within research community, but they also strive for grants, for projects, for venues of publication. Researchers must interject their knowledge in the "common pot" and give up their intellectual property rights to help the collective knowledge, and they must select carefully where to apply their time and efforts. Loss of money and time can be overcome by using the open data generated by others; it would fasten the process of data collection and analysis, and will save the amount to be spent on equipment, publication fees, or other research necessities.

Sustainable release of data and maintaining open data infrastructure seems to be costly, even if the research funding is used for open data sharing, but the overall benefit may substantially decrease the cost of doing research. Data release is more effective if the data are curated in many ways which will make it useful to others over some long period of time. Similarly, more needs to be known about potential uses and users of research data, following framework can be implemented. A central repository and open data infrastructure framework are very much required which can cater to requirements of entire ecosystem of open data. Framework can help all the stakeholders to collaborate, share and use the data; it should be designed to cater to variety of requirement of different stakeholders to enhancing the use of datasets in different ways.

Open data infrastructure must be rooted in a clear policy framework to make sure all data is accessible to all in a standard manner, while defining policy framework is important, it is also essential to establish a culture of openness. Proper collaborative infrastructure can enable open data to go beyond the current level of data availability for researchers, and community engagement could result in an unremitting dialogue between all the stakeholders; to foster innovation and socioeconomic growth, a collaborative framework of open data infrastructure has been described in Fig. 3.

The identified infrastructure and its associated elements provide a glimpse of how to implement such infrastructure, where not only the data but underlying model and research techniques would also be stored. Further research in this domain is required to refine the reference architecture and its associated elements and modules. From the process perspective, there is a strong need of a working system backed by robust policy and technological framework, where all stakeholders, i.e., researchers, governments, developers, activist, public, can join their hands to form a federated data infrastructure, not only "for the end-users," but "with the end-users." Implantation of open data infrastructure is the only way to gain

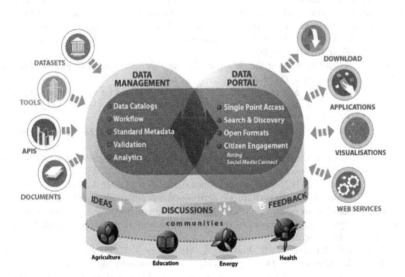

Fig. 3 Framework of open data infrastructure

confidence and acceptance of user communities, and to take the open data proposition into broader scope of innovations and socioeconomic growth.

References

1. Bechhofer, S., De Roure, D., Gamble, M., Goble, C., & Buchan, I. (2010). Research objects: Towards exchange and reuse of digital knowledge.
2. Borgman, C. L. (2012). The conundrum of sharing research data. *Journal of the Association for Information Science and Technology, 63*(6), 1059–1078.
3. Boulton, G. (2014). The open data imperative. *Insights, 27*(2).
4. Bowers, S., McPhillips, T., Wu, M., & Ludäscher, B. (2007, June). Project histories: Managing data provenance across collection-oriented scientific workflow runs. In *International Conference on Data Integration in the Life Sciences* (pp. 122–138). Berlin: Springer.
5. Buddenbohm, S., Cretin, N., Dijk, E., Gaiffe, B., De Jong, M., & Minel, J. L., et al. (2016). State of the art report on open access publishing of research data in the humanities. Doctoral dissertation, DARIAH.
6. Childs, S., McLeod, J., Lomas, E., & Cook, G. (2014). Opening research data: Issues and opportunities. *Records Management Journal, 24*(2), 142–162.
7. Conradie, P., & Choenni, S. (2012, October). Exploring process barriers to release public sector information in local government. In *Proceedings of the 6th International Conference on Theory and Practice of Electronic Governance* (pp. 5–13). ACM.
8. Davies, T., & Frank, M. (2013, May). 'There's no such thing as raw data': Exploring the socio-technical life of a government dataset. In *Proceedings of the 5th Annual ACM Web Science Conference* (pp. 75–78). ACM.
9. Estermann, B. (2014). Diffusion of open data and crowdsourcing among heritage institutions: Results of a pilot survey in Switzerland. *Journal of theoretical and applied electronic commerce research, 9*(3), 15–31.
10. Fox, X. M. P., Beaulieu, S. E., Fu, L., Di Stefano, M., & West, P. (2016). Documenting provenance for reproducible marine ecosystem assessment in open science. *Oceanographic and Marine Cross-Domain Data Management for Sustainable Development*, 100.
11. Gurstein, M. B. (2011). Open data: Empowering the empowered or effective data use for everyone? *First Monday, 16*(2).
12. Hester, J. R. (2014). Closing the data gap: Creating an open data environment. *Radiation Physics and Chemistry, 95*, 59–61.
13. Himmelreicher, R. K., & Stegmann, M. (2008). New possibilities for socio-economic research through longitudinal data from the research data centre of the German federal pension insurance (FDZ-RV). *Schmollers Jahrbuch, 128*(4), 647–660.
14. Hrynaszkiewicz, I. (2011). The need and drive for open data in biomedical publishing. *Serials, 24*(1).
15. Janssen, K. (2011). The influence of the PSI directive on open government data: An overview of recent developments. *Government Information Quarterly, 28*(4), 446–456.
16. Janssen, S., Porter, C. H., Moore, A. D., Athanasiadis, I. N., Foster, I., & Jones, J. W., et al. (2015). Towards a new generation of agricultural system models, data, and knowledge products: Building an open web-based approach to agricultural data, system modeling and decision support. AgMIP. *Towards a New Generation of Agricultural System Models, Data, and Knowledge Products*, 91.
17. Kidwell, M. C., Lazarević, L. B., Baranski, E., Hardwicke, T. E., Piechowski, S., Falkenberg, L. S., … & Errington, T. M. (2016). Badges to acknowledge open practices: A simple, low-cost, effective method for increasing transparency. *PLoS Biology, 14*(5), e1002456.

18. Knowledge Exchange Homepage, http://www.knowledge-exchange.info/, last accessed 2017/08/31.

19. Lenhart, A., Simon, M., & Graziano, M. (2001). The Internet and Education: Findings of the Pew Internet & American Life Project.

20. Mishra, A., Misra, D. P., Kar, A. K., Babbar, S., & Biswas, S. (2017, November). Assessment of open government data initiative—a perception driven approach. In *Conference on e-Business, e-Services and e-Society* (pp. 159–171). Springer, Cham.

21. Missier, P., Ludäscher, B., Bowers, S., Dey, S., Sarkar, A., Shrestha, B., ... & Goble, C. (2010, November). Linking multiple workflow provenance traces for interoperable collaborative science. In 5th Workshop on Workflows in Support of Large-Scale Science (WORKS) (pp. 1–8). IEEE.

22. Moore, S. (2014). *Issues in Open Research Data* (p. 164). Ubiquity Press.

23. Nogueras-Iso, J., Zarazaga-Soria, F. J., Lacasta, J., Béjar, R., & Muro-Medrano, P. R. (2004). Metadata standard interoperability: Application in the geographic information domain. *Computers, Environment and Urban Systems, 28*(6), 611–634.

24. Nosek, B. A., Alter, G., Banks, G. C., Borsboom, D., Bowman, S. D., Breckler, S. J., ... & Contestabile, M. (2015). Promoting an open research culture. *Science, 348*(6242), 1422–1425.

25. OAPEN Homepage, http://www.oapen.org/home, last accessed 2017/08/31.

26. Open Data Institute Homepage, http://theodi.org/, last accessed 2017/08/31.

27. Open Knowledge Foundation Homepage, https://okfn.org/, last accessed 2017/08/31.

28. OpenAIRE Homepage, https://www.openaire.eu/, last accessed 2017/08/31.

29. Reichman, O. J., Jones, M. B., & Schildhauer, M. P. (2011). Challenges and opportunities of open data in ecology. *Science, 331*(6018), 703–705.

30. Sawicki, D. S., & Craig, W. J. (1996). The democratization of data: Bridging the gap for community groups. *Journal of the American Planning Association, 62*(4), 512–523.

31. Starr, J., Castro, E., Crosas, M., Dumontier, M., Downs, R. R., Duerr, R., ... & Hourclé, J. (2015). Achieving human and machine accessibility of cited data in scholarly publications. *PeerJ Computer Science, 1*, e1.

32. Stevens, B. J. (1984). *Nursing theory. Analysis, application, evaluation* (2nd ed.). Boston: Little, Brown.

33. Tananbaum, G. (2008). Adventures in open data. *Learned Publishing, 21*(2), 154–156.   .

34. Verma, N. (2013, August). Open data for inclusive governance. In Joint Proceedings of the Workshop on AI Problems and Approaches for Intelligent Environments and Workshop on Semantic Cities (pp. 5–5). ACM.

35. Verma, N., & Gupta, M. P. (2012). *Open government data: More than eighty formats.* Paper presented at the 9th International Conference on E-Governance (ICEG 2012), Cochin, Kerala, India.

36. Verma, N., & Gupta, M. P. (2013, October). Open government data: Beyond policy & portal, a study in Indian context. In *Proceedings of the 7th International Conference on Theory and Practice of Electronic Governance* (pp. 338–341). ACM.

37. Verma, N., & Gupta, M. P. (2015, November). Challenges in publishing Open Government Data: A study in Indian context. In *Proceedings of the 2015 2nd International Conference on Electronic Governance and Open Society: Challenges in Eurasia* (pp. 1–9). ACM.

Managing Research Data by R&D Community in Nuclear Data Science in India

S. Ganesan

1 Introduction

The author believes that an increased use of big data science and use of information technology in India will positively contribute to the acceleration of human progress and help reduce the gap in human development index between India and other developed countries. The use of big scientific databases helps us do better science as knowledge progresses with increasing complexities in data generation by costly and improved micro- (i.e., differential) and macro (i.e., integral)-experiments. Big data science is a natural consequence of advances in computer hardware and software, in experimental techniques, analysis, and in multiscale, multiphysics modelling (MMM3) simulations. The use of big data science increases system intelligence and system performance bringing in factors of correlations. Generation and use of big numerical databases form part of MMM3 in implementing Bhabha's 3-stage programme. See http://www.dae.gov.in and links to various institutes therein.

This write-up deals with the perspectives in managing scientific R&D databases taking author's experience, as an example in the context of Indian nuclear scenario, in the field of nuclear data science. The write-up does not attempt to cover all aspects of nuclear data science, to save space.

The R&D scientific databases that we deal with in this article are specifically the numerical data available in the website http://www-nds.indcentre.org.in, the India mirror of the website in Vienna, viz., http://www-nds.iaea.org of the International Atomic Energy Agency (IAEA). Such numerical values of nuclear physics data are the results of basic nuclear physics experiments and thus are "constants" of nature describing the nuclear structure and those characterizing the interaction probability between two nuclear systems. Physics as a science is based upon observations,

S. Ganesan (✉)
Raja Ramanna Fellow of the DAE, Bhabha Atomic Research Centre,
Trombay, Mumbai 400085, India
e-mail: ganesan555@gmail.com

© Springer Nature Singapore Pte Ltd. 2018   45
U. M. Munshi and N. Verma (eds.), *Data Science Landscape*,
Studies in Big Data 38, https://doi.org/10.1007/978-981-10-7515-5_3

called "scientific experiments" in nature. Mathematical modelling of physical events needs a quantitative expression of scientific experiments (observations) in the form of numerical scientific data.

The numerical values of data characterizing the physical outcome of events in experiments are the quantitative results of experiments. These numbers are meaningful only if full descriptions of uncertainties in terms of all partial uncertainties and their correlations are specified by the experimenter. Quality-assured and complete documentation including the full numerical data along with complete error specifications is essential to preserve the valuable experience for nuclear applications towards the benefit of mankind [1].

At the outset, it may be noted that the scientific and engineering databases that we deal with are not bibliographic databases, such as the International Nuclear Information System (INIS), of the IAEA (Ref: https://www.iaea.org/inis/). The scientific databases that we deal with in simulations consist of the voluminous amount of numerical data, in some cases, running to hundreds of gigabytes at a raw level, processed level and recommended level. The voluminous numerical data produced in scientific experiments are thus "signatures" of Goddess of Knowledge, Saraswati and are not to be confused with the traditional and computerized bibliographic databases, such as the INIS, which are highly useful and have their merit in their own way.

2 Applications of Nuclear Data

Development of knowledge base, providing an accurate description of basic nuclear interactions, is a fundamental and natural part of the evolution of nuclear science and technology. Applications of nuclear databases include all areas of nuclear science and technology, covering energy applications (fission reactor design; nuclear fuel cycles; nuclear safety; reactor monitoring and fluence determination; waste disposal and transmutation; accelerator driven systems; fusion device design and plasma processing technologies) as well as non-energy applications (cancer radiotherapy; production of radioisotopes for medical and industrial applications; personnel dosimetry and radiation safety; nuclear safeguards; environmental monitoring and clean-up; materials analysis and process control; radiation damage studies; detection of concealed explosives and illegal drugs; exploration for oil and other minerals) and basic research (e.g. nuclear astrophysics and human curiosity-based efforts towards understanding the origin of the Universe) and human resource development. Nuclear data science is a front line field including cutting-edge technologies in basic data measurements and analysis helping to provide a better scientific basis for nuclear applications. Obviously, new concepts of reactor designs will have a sound scientific basis if the nuclear data used are accurate.

Nuclear systems for energy and non-energy applications have come to stay in the history of mankind. Relative to the history of mankind, the age of man-made nuclear energy, about seven decades old, is indeed very much in its infancy. In the opinion of the author, nuclear energy itself has been introduced to mankind without adequate knowledge of nuclear data physics that would be demanded by advanced reactor designs. This statement on inadequacy of nuclear data may sound strange with hundreds of Gen-I and GEN-II reactors operating today, since the 50s, but it must be stressed that the safety and operational requirements of existing nuclear power plants have all been well engineered with a number of one-to-one mock-up experiments providing adequate and perceived conservative safety margins, with adjustment of the nuclear data to fit the particular range of integral experiments.

Design of innovative reactors is shaped by a number of issues and considerations, such as materials development, engineering viability and passive safety inclusions. One of the important points is that the basic nuclear data physics research and associated data science have been essential in shaping concepts of nuclear power reactor designs.

India recognizes that nuclear data science is an essential base technology effort as part of big data science in meeting all the Indian nuclear data needs of different types of nuclear reactors such as thermal, fast, fission-fusion and accelerator-driven sub-critical systems, which involve multiple fuel cycles (U-Pu and Th-U). India follows closed nuclear fuel cycles. It is recognized that long-term nuclear data needs of closed fuel cycles with multiple fuels (U-Pu and Th-U), with high burnup, are demanding in the Indian context of Bhabha's 3-stage nuclear power programme. The author believes that, internationally, the best nuclear reactor design that would not even be remotely accident prone during the entire fuel cycle that would be with minimum radioactive waste, fully proliferation resistant, with maximum tolerance of normal and even of remotely possible operator errors, man-made and natural disasters are yet to be made. In the humble opinion of the author, the incident in the nuclear industry such as Fukushima could have been avoided if big data science approach had been rigorously followed, because, following an approach of big data science would have included, by design, a worldwide and comprehensive expertise to make reactors much safer.

3 Multiphysics Multiscale Modelling (MMM3)

MMM3 is the most complete and detailed form of modelling in 3-Dimensions, for a computerized and automatic coupling of various physical processes, with associated big data science efforts in each of them, such as the following:

- Neutronics of steady state and transients,
- thermal-hydraulics of core, full plant, one- and two-phase flows,
- structural mechanics of the fuel and other parts,
- radiation-induced damage, burnup evolution,
- chemical effects,
- radio-toxicity for each of the component overall operating phase-space,

- decommissioning,
- waste disposal.

These physical processes are coupled and extend over large time scales varying from picoseconds to centuries. Thus, a large number of numerical and physical models are employed and validated at various levels with supporting experiments, wherever possible. Data uncertainties coupled with model sensitivity studies will further identify areas where improvements are needed in differential and integral data demanding specific experiments and/or improvement of basic physics data and models.

The obvious advantage if one is able to make progress in MMM3 is that a number of costly integral experiments that would otherwise require several years to conduct can be significantly reduced. A successful programme of MMM3 in current and future nuclear reactors involves a well coordinated scientific teamwork with several disciplines participating over a long term. The culture of the teamwork to effectively make progress has to be nurtured. The evolution of this strategy also involves fixing a number of basic physics data that goes in the neutronics, thermal hydraulics, radiation damage, chemical changes etc., with reasonable accuracy and performing a large number of coupled sensitivity studies to identify areas and needs of experiments, both basic and applied, in each of the disciplines. The resulting design document of a plant using a perfect MMM3 with the bid data science approach is a dream come true for any plant operator.

Nuclear data physics efforts are base technology efforts and are an essential part of MMM3 of nuclear systems for all energy and non-energy applications. The big data science is not just about the matrix of voluminous multiparameter data generated in experiments. The efforts in data science help to quantitatively model the system with all detailed data and help validation of models at differential and at limited to extended integral levels. This has become possible in the last two decades because of rapid improvements in information technology, both computer hardware and software. Also, massive parallelization has added confidence to this initiative. A number of papers and documents are available on the Internet on the topic of MMM3.

The area of MMM3 is relatively a very recent development that is taking place in many branches of science and technology. Some aspects of MMM3 for nuclear reactor design studies are available for the interested reader in the website, for instance, in http://www.cea.fr/english/Pages/resources/clefs-cea/big-data-challenge.aspx.

MMM3 attempts to fulfil the long-standing requirement that explicitly take into account, when modelling an advanced nuclear reactor system in detail for which the behaviour is to be predicted for all processes (neutronic, thermal-hydraulic, structural mechanics of the fuel and other parts, radiation-induced damage, burnup evolution, microstructural characteristics such as porosity, cluster size and distribution, grain size and chemical effects) simultaneously.

The MMM3 makes the full use of all of the relevant knowledge in the form of detailed numerical data quantifying the basic physics processes. These physics and engineering databases are large in size and typically run into Gbytes posing challenges in terms of quality assurance to interface them with the codes respecting the physics laws for their condensation in some cases. The use of Monte Carlo codes and parallel algorithms with current computing power and resources makes it possible to attempt full reactor core calculations over several recycling of nuclear fuel.

The MMM3 takes into account the strong couplings arising between basic processes of diverse nature over several orders of magnitude of time scales. In studies of reactor physics of current and advanced reactors, for instance, strong coupling occurs between structural mechanics, neutronics and thermal hydraulics. In the area of material science, for instance, in the case of simulation of reactor pressure vessel (RPV) integrity, the motivation is to obtain the integrity limits of RPV in the entire lifetime of several decades of the operating power. In this case, the MMM3 starts from a simulation at the most fundamental microscopic atomic scale to simulate quantities such as the creation of frenkel pairs, defects migration and evolution, ab initio molecular dynamics and dislocation dynamics. These phenomena are tracked over several decades and lead to the prediction of change in the strength and integrity of the RPV.

The challenging issue in MMM3 is that the correct physics information should pass from one scale to the next with full consistency of the physical laws and with no break in continuity. The MMM3 should enable us to "zoom in" on regions that are particularly sensitive to certain parameters, such as nuclear plant performance, energy growth potential or system integrity affecting the structure.

The confidence to take up MMM3 has arisen in the history of nuclear energy and other areas because of improved computer resources and software developments coupled with the ability to continuously update the basic physics and engineering databases to unprecedented details and accuracy and integral databases at different integral levels. Many of the effects in operating environment need extrapolation by theoretical means based upon physics laws from well-defined basic integral experiments. For instance, taking the relatively known case of a physics reactor critical facility, one may measure the coolant void reactivity effect at cold temperature in a zero power, one-to-one engineering mock-up critical facility but verification of the void effect at high burnups and higher temperatures encountered in normal operation can be obtained only through reliable modelling and calculations, which require a firm scientific basis for extrapolation to operating conditions. Estimation of the scenario and effects of unintended over-power transients for which we may not even be able to do an experiment would greatly benefit from a reliable software that performs MMM3.

The MMM3 efforts also pose challenges in terms of development of front-end and post-processors and visualization tools and efficient message passing between segments that are coupled. The problem of storage and coupling is demanding. Challenges exist when sensitivity studies are required to be made in 3-Dimensions and directly compatible with (computer aided design) AUTOCAD drawings. Front-end techniques such as computational fluid dynamics (CFD) have to be mastered for a variety of problems arising in MMM3. Benchmark experiments have to be ongoing and conducted for each segment and each level of sophistication to bring the efforts to a credible level. Getting a database of experimental benchmarks to validate MMM3 at micro- and macro-integral levels is important. For the full scale of MMM3 performing and making quality experimental benchmarks is a challenging task.

4 Nuclear Data Science as Part of Big Data Science

Design of advanced nuclear systems demands an assessment of confidence margins in plant's design parameters. The errors in the design parameters of the advanced nuclear systems due to errors arising from the uncertainties in basic nuclear data are addressed by the nuclear community by and large through the covariance methodology. In these cases of usage of covariance methodology, it is assumed that the specification of basic cross section data is determined completely by the mean and covariances that are available in the basic evaluated nuclear data files. India has initiated an interesting programme on nuclear data covariances [1].

Internationally, the evolution of the subject of big scientific databases and knowledge management has broadly included several identified base technology efforts mentioned below, in a generic sense across disciplines, using advances in computer science and numerical simulations, physics and technology. Generically, in the Indian context, the following stages of activities are being nurtured. The steps in nuclear data science mentioned below are also common to other disciplines, and thus the philosophy and techniques can generically apply across disciplines.

4.1 Measurements and Creation of "Raw Data"

Physics is observational. Healthy citation practices of journal articles containing details of experiments and laboratory reports are being followed in nuclear data science. Indian leadership recognizes that experimental studies to generate nuclear data of high quality require good quality research facilities to determine nuclear cross sections covering neutron and charged-particle reactions, and nuclear structure and decay data (all with well-defined uncertainties and to high accuracy), with the ability to cover the nuclear data physics needs for advanced fission and fusion systems (that also has thorium fuel and closed fuel cycle), analytical science and nuclear medicine. This involves, because of the need for neutron sources and particle accelerators, significant costs and capital outlay. All nuclear physicists, by default, would provide the expertise to undertake facility development tailored to nuclear data physics measurements. For the same nuclear reaction, experiments are encouraged to be re-done. Re-measuring the basic nuclear data with better accuracy in a given energy range and/or covering new energy regions where data have not been measured previously is recommended and funding provided as the target accuracy for many nuclear applications have not been met. Literature has plenty of examples of the same physical quantity having been re-measured and analysed more accurately, when any of the features of the experimental set-up such as better neutron sources, better energy resolution, higher intensity of neutrons, better enriched and purer samples and with better detectors as and when become available by developments.

4.2 Systematic Compilations of Raw Experimental Data

EXFOR [Exchange Format (for experimental nuclear reaction data)] is the collection raw experimental data [2]. See http://www-nds.indcentre.org.in/exfor/exfor. htm. Performing an experiment is very expensive in terms of equipment and resources of expert manpower. Preserving the experimental details is important. No judgement on the quality of the data generated in the experiment is attempted, by design, during this EXFOR compilation process. The EXFOR compilation effort requires deep technical knowledge of nuclear physics to ensure quality in EXFOR compilations. Challenges exist in promoting experimenter's documentation of partial errors and their correlations to enable compilers include the error specifications data properly in EXFOR compilations.

Compilation of raw experimental nuclear physics data into EXFOR is generically a challenge as in many other subject areas. The compiler needs to have an in-depth domain expertise and an attitude to perform the compilation task. Authors of compilation work are not always recognized by scholarly citations to compilation efforts. There are challenges in man-power training, attitude problems and in managing this task in terms of funding and in finding people who are willing to commit to this compilation work over a long term. In order to solve these problems, EXFOR workshops in India have been evolved as a new and unique managerial initiative. These Indian EXFOR workshops have been phenomenally successful. Introduction of EXFOR culture in people including in basic nuclear physics has become relatively an easier task with the new managerial initiatives of holding EXFOR workshops in India. Through the EXFOR workshops, the Nuclear Data Physics Centre of India (NDPCI) has been phenomenally successful to bring people in various fields, e.g. nuclear physics, reactor and radiochemistry divisions, covering experimentalists and theoreticians of Bhabha Atomic Research Centre (BARC), Mumbai, Indira Gandhi Centre for Atomic Research (IGCAR), Kalpakkam, Variable Energy Cyclotron Centre (VECC), Kolkata, plant reactor physicists from Nuclear Power Corporation of India (NPCIL), and students and faculty from various universities in different disciplines from across India. Where nuclear data are not provided, and publication contains graphs of data, the graphs are digitized to include the data in the EXFOR database, with the approval of the authors, wherever possible. The EXFOR compilers are required, by design, to interact patiently with the authors. The experimenter compiling the EXFOR himself/herself will be the ideal win–win situation. In India, slowly, many experimenters are understanding and increasingly supporting the EXFOR compilation activity to compile own experiments as a result of the workshops.

In the opinion of the author, there is a national need, in the Indian context, for funding agencies to insist that research authors take interest and responsibility to submit numerical data generated in experiments for inclusion in EXFOR database, at the time of acceptance by a journal for publication, as a condition for publication and further funding. This will increase the speed and efficiency in the EXFOR compilation activities.

Use of tools used in a compilation such as digitizers require training. Evolution of common formats (XML) needs considerable development work. The EXFOR compilation activities are part of classical data physics activities and have made a phenomenally successful start, since 2006, in India. Indian EXFOR workshops are serving as a role model that can be tried as a follow-up in other areas of data science in data compilation efforts. Such data compilation efforts have to be a long-term, ongoing, national activity in each area of the topic of data science.

4.3   Evaluation of Basic Nuclear Data

The phrase "Evaluation of nuclear data" is not understood by many in India and is sometimes mistaken for nuclear model-based calculations of numerical nuclear data. The nuclear data evaluation process includes re-normalization of available experimental data in EXFOR, science-based models, systematics and statistical/ mathematical inference tools. The process of evaluations of basic nuclear data results in numerical values of evaluated mean values and covariances, which include nuclear model-based predictions through the use of inference tools, such as the Bayesian approach. Use of advanced statistical tools is part of this task. Documentation for training purposes on how an evaluation was performed is historically missing, internationally, in many cases. The newcomers do not readily have benchmarks in evaluations to learn. The results of evaluation by different countries for a given nuclear physics data have not yet universally converged.

In the Indian context, as nuclear data science got importance since 2004, the task of evaluation of nuclear data has been recently initiated in India, but we need to gain a lot of experience, developing expertise, before being able to make high quality basic nuclear data evaluations.

The research in frontier areas of basic nuclear data physics, associated knowledge management and critical evaluation of associated uncertainties are being rigorously sustained, well supported and pursued further in BARC. The existing strength of currently available, state-of-the-art nuclear databases in use for various applications is highly commendable but inadequate to meet the nuclear data needs of new reactor concepts as different neutron energy spectra, materials and compositions are involved. Thus, nurturing indigenous evaluation of basic nuclear data is essential.

4.4 Visualization and Processing of Large Databases for Monte Carlo Applications and Processing and Production of "Plug-in" Nuclear Data Libraries

Physics laws based nuclear data processing of basic evaluated nuclear data files has also to be performed using complex processing codes in order to generate problem dependent multienergy group data values. The processing of basic data files is also performed to generate compact energy pointwise nuclear data libraries for use in Monte Carlo applications. The processed libraries, i.e. the "plug-in" nuclear data libraries, were mainly seen, until a few years ago, by reactor physicists as "the nuclear data tables" used in the design. The textbooks in reactor physics do not generally touch on the topic of nuclear data processing, which therefore is taken for granted. Also, it is well recognized that neutronic Monte Carlo codes can be developed indigenously only when formats and conventions of large nuclear databases such as ENDF/B can be digested and interfaced with Monte Carlo simulations. The nuclear data processing tasks are generally performed by nuclear data processing specialists in order to produce plug-in nuclear data files starting from basic evaluated nuclear data files. The step of nuclear data processing is essential as the basic evaluated data in ENDF/B format cannot be directly coupled to application software. Scholarly citations of work on processing of nuclear data are relatively low in the field.

Expertise in processing evaluated nuclear data files will have to be kept nurtured in the Indian context. Ready plug-in nuclear data libraries available from other international sources are not sufficient to meet all our needs of advanced reactors using multiple fuels with long burnup and with the closing of the fuel cycles.

4.5 Design Calculations Using Plug-in Nuclear Data Files Obtained in Step 4

In this step, one starts from the "plug-in" nuclear data generated in Step 4.4. The innovative design of the complex nuclear system such as a nuclear reactor starting from plug-in nuclear data tables is in itself a complex scientific and engineering task. Applications using plug-in data libraries, e.g. for MMM3 and development of the associated complex computer software packages are part of this task. This area of using application software towards the design of nuclear systems is well cited depending on quality and impact.

Historically, reactor physics studies in India started from this Step 4.5, i.e. from the point of using ready "plug-in" (and processed) nuclear data libraries from foreign sources. Fortunately, the Indian nuclear data physics activities supported by the Department of Atomic Energy-Board of research in Nuclear Sciences (DAE-BRNS) in the last 12 years have expanded considerably beyond this perspective to initiate and include R&D activities on our own nuclear data measurements, compilations, evaluation, processing and integration testing. Many of the concepts such as "evaluation of basic data", "EXFOR compilations" and Integral criticality benchmarking" are new initiatives, since 2004, in the Indian context.

4.6 Integral Results Based upon Physical Real System Experiments

Using intelligently, the available recommended "plug-in" nuclear data to perform tasks mentioned in Step 4.5 and Step 4.6 require in itself a huge effort. However, what is not generally stated in textbooks is that the scientific and base technology efforts (Steps 4.1–4.4) in order to produce ready "plug-in" data (Step 4.5) starting from "no raw data available" status and going through Steps 4.1–4.4 are three to four orders of magnitude more than the efforts made from the stage (Step 4.5) of using available recommended data in a plug-in form to perform innovative design of complex systems.

Experimental basic nuclear data physics measurements using accelerator and reactor-based neutron sources and also a programme of the critical facility for integral validation of reactor physics data of Advanced Heavy Water Reactor (AHWR) at BARC (www.barc.gov.in) have been in good progress in BARC/DAE.

The integral validation task involves performing a number of sensitivity studies, assessment of uncertainties in system characterization and benchmarking to match, for instance, the QA of the International Criticality Benchmark Evaluation Project of the US-DOE, which provides a database of criticality benchmarks created by several countries including India. See the ICSBEP website: http://icsbep.inl.gov for more details.

The above-mentioned organization of nuclear data science (Steps 4.1–4.6) efforts has helped immensely in eliminating the pitfalls associated with individual's limited efforts to knowledge manage the entire knowledge base related to a specific study/application from the voluminous microscopic data to integral data.

5 Citation Practices and Guidelines in Nuclear Data Science

It may be pointed out that nuclear data scientists generally follow the citation guidelines very well. The guidelines for citations are provided in websites and dataset related documentation. In 1996, Mclane [3], for instance, has provided guidelines for citations in the nuclear data field. As pointed out by Mclane:

"Data obtained from the databases residing at the member organizations of the Nuclear Data Centers Network should be properly cited. In general, there should be a citation of the original source of the information used, as well as of the database from which the data were extracted. The source of the information should be cited as:

- Data retrieved (or extracted) from the (center name) Online Data Service,
- Data retrieved (or extracted) from the (center name) WorldWideWeb site,
- Data received by electronic file transfer from (center name),
- Data received from (center name)

These data bases may contain essential information which does not exist in a published article.

Since the databases are periodically updated, it is important to include the date and/or revision number of the version of the database used."

We also see that in order to encourage and help the citation efforts in the field of data science, clear instructions on how to cite a given data science work are being given by data scientists. For instance, the Japanese nuclear data evaluation centre clearly provides guidelines in their data centre website on how to cite the evaluated nuclear data file, version 4.0 of the Japanese evaluated nuclear data file, JENDL-4.0, in the website providing the data: http://www.ndc.jaea.go.jp/jendl/j40/j40.html.

N. Otuka has noted [4] the following in an email (Nov. 2015) to the author of this write-up, that regarding citation of experimental data sets, EXFOR users usually cite papers where the experimental data sets are published, as is of course mandatory. He notes that, however, the effort of EXFOR data compilers is not visible at all in general. The EXFOR users are encouraged to cite, for instance, Ref. [2] as mentioned in the EXFOR website: http://www-nds.indcentre.org.in/exfor/exfor.htm.

Otuka also notes in his recent email that the situation is opposite for (Evaluated Nuclear Structure Data File) ENSDF evaluations. "People often access them through web interfaces (e.g. NuDat, Livechart of Nuclides) and they cite only NuDat without citing original evaluation reports. The data users and collaborators are encouraged to cite the original ENSDF related evaluation reports published in Nuclear Data Sheets".

There are nice examples of credits to datasets in GEANT4, which is a software package for the simulation of the passage of particles through matter. The website of GEANT4 acknowledges the use of external data libraries (http://geant4.cern.ch/support/datafiles_origin.shtml) in compiling the GEANT4 data libraries. Information on these sources is currently provided also in their publications, including in the journal reference NIM A 506 (2003), 250.

There are special cases where the users of nuclear data and designers of reactors do not follow the citation practices with respect to used sources of nuclear data, though the situation is improving. Citations to plug-in libraries and efforts (Step 4.4) are not done in some rare cases by reactor designers in India. In some cases, historically, in the Indian context, the reactor physics designers, who are highly competent, have felt that reference in the documentation/journal article to nuclear data sets used in design might affect the belief in and follow-up of the calculated design numbers. For the same reason and for simplicity, nuclear data sensitivity studies were not always performed. Managing the R&D in nuclear data science and encouraging practices of scholarly citations, in the Indian context, to data sets are now being encouraged.

6   Concluding Remarks

The data science activities in nuclear data science tailored to the Indian Atomic Energy applications are generically useful in other areas of sciences, in the Indian context. In the Indian context, there is a need to horizontally share the data science expertise across various disciplines. Data compilations workshops are needed to be encouraged in every discipline. The talk is on numerical databases, citation needs and procedures taking nuclear data science R&D as an example. The citation protocols are well defined, and Indian scientists and engineers are being kept sensitized.

In this paper, we dealt with some aspects of nuclear data physics requirements with respect to the development of advanced and new reactor concepts. India recognizes the need for reliable nuclear data for all evaluations for several hundreds of isotopes/elements in all stages of the nuclear fuel cycle. In the nuclear field, the role of multiphysics multiscale modelling with "big" data science including detailed nuclear data cannot be overemphasized in attempts to evolve the best reactor design.

India has been internationally acclaimed by the International Nuclear Reaction Data Centre (NRDC) network for its new managerial initiatives in organizing EXFOR workshops and making, thus far, more than 345 EXFOR entries (as on 13 April 2017) based upon published Indian nuclear physics experiments (see https://www-nds.iaea.org/exfor-master/x4compil/exfor_input.htm) at the time of writing this paper. India also has been contributing to Indian experimental criticality reactor benchmarks of the ICSBEP efforts (http://icsbep.inl.gov). India is also participating in the international collaborations of CERN n_TOF programme (https://ntof-exp.web.cern.ch/ntof-exp/). These new initiatives which are milestone development activities have indeed provided increased visibility to India's nuclear physics activities in the world in a highly positive manner. India has also successfully initiated a programme on the characterization of uncertainties in nuclear data and propagation of uncertainties in the form of covariances in nuclear data [1]. These new nuclear data science activities have been a significant value addition to basic and applied nuclear physics activities in the country.

As a general remark, in the opinion of the author, the use of big data science in India will help accelerate progress to reduce the gap in human development index between India and other developed countries.

Acknowledgements The author wishes to express his thanks to Dr. Usha Mujoo Munshi, Prof. Krishan Lal and the organizers for the invitation to deliver this talk. This brief write-up corresponds to the Powerpoint slides used in the invited talk in the International Workshop on "Big and

Open Data: Evolving Data Science Standards & Citation Attribution Practices", organised by: Indian Institute of Public Administration, Indian National Science Academy & INFLIBNET and JNU (Knowledge Partners), 5–6 November 2015 at the venue: Indian National Science Academy, Bahadur Shah Zafar Marg, New Delhi, PIN: 110002, India.

References[1]

1. Ganesan, S. (2015). Nuclear data covariances in the Indian context-progress, challenges, excitement and perspectives. *Nuclear Data Sheets, 123,* 21–26. (and references therein).
2. Otuka, N., Dupont, E., Semkova, V., Pritychenko, B., Blokhin, A. I., Aikawa, M., et al. (2014). Towards a More Complete and Accurate Experimental Nuclear Reaction Data Library (EXFOR): International Collaboration Between Nuclear Reaction Data Centres (NRDC). *Nuclear Data Sheets, 120,* 272.
3. McLane, V. (1996). *Citation guidelines for nuclear data retrieved from databases resident at the "Nuclear data center's network".* Report: BNL-NCS-63381, BNL, USA. See: www.iaea.org/inis/collection/NCLCollectionStore/_Public/28/015/28015549.pdf.
4. Otuka, N. (2015). NDS, IAEA, Email to Ganesan, dated October 3, 2015.

---

[1]Given 3 references are illustrative, to save space. References in these 3 references are useful. Additionally, weblinks have also been provided in the text for the interested reader.

Big Data in Astronomy and Beyond

Ajit Kembhavi

1 Astronomical Observations

Modern astronomical research spans a very broad range of the electromagnetic spectrum, from the radio to infrared, optical, ultraviolet, X-ray and gamma-ray regions. The observations in these regions are made from ground- and space-based telescopes. Data are obtained from observations of individual or groups of objects, as well as from surveys of large areas of the sky, which provide data on billions of objects.

Observations in different ranges of the electromagnetic spectrum require quite different types of telescopes. In the region of visible light, which covers the part of the spectrum that we can see with our eyes, observations are made with optical telescopes which use appropriately shaped mirrors to gather light and to focus it. Such telescopes are mostly located on the ground, with the largest of these at the present time having a mirror with a diameter of about 10 m. There are also optical telescopes in space, like the Hubble Space Telescope. Radio telescopes, which are used to observe emission from celestial objects in the radio part of spectrum, can consist of a single very large dish or a number of smaller dishes spread over tens of kilometres or even at intercontinental distances. X-ray telescopes have to be located on artificial satellites orbiting the earth outside the atmosphere, since X-rays from celestial sources passing through the atmosphere are completely absorbed by it.

All these observations generate large volumes of data, particularly when telescopes are used to conduct surveys of the sky. In these surveys, either the whole sky or large portions of it are observed, thus recording the emission from millions or even billions of objects and any background emission which is present, generating very large volumes of data over the period of the survey, which can stretch to several years.

A. Kembhavi (✉)
Inter-University Centre for Astronomy and Astrophysics, Pune, India
e-mail: akk@iucaa.in

© Springer Nature Singapore Pte Ltd. 2018
U. M. Munshi and N. Verma (eds.), *Data Science Landscape*,
Studies in Big Data 38, https://doi.org/10.1007/978-981-10-7515-5_4

In recent times, surveys of selected areas of the sky are being carried out repeatedly, at cadences which may vary from minutes to hours to days to months, so that changes in celestial objects over time can be studied. The most important data obtained from such surveys is on transients, which are objects in the sky which change in brightness over short periods of time. These could be objects in our own galaxy, like cataclysmic variables or supernova explosions, or very bright supernovae or gamma-ray bursts in distant galaxies, or even completely unknown kinds of objects being detected for the first time. These surveys generate very large volumes of data every night. It is necessary to quickly establish the identity of any transient which is found, based on the limited data which may be available on any object, so that the object can be observed with suitable telescopes before it fades from view. This brings in machine learning, which uses statistics- and computer science-assisted supervised and unsupervised software techniques to make intelligent and informed guesses.

2 Big Data in Astronomy

Astronomy is a science based on the analysis of observations. In ancient times, the data was collected with the human eye. This continued to the time of Tycho Brahe in the late sixteenth century and beyond, until the first use of a small telescope for astronomical observations by Galileo Galilei in 1609. The data collected and recorded were modest in volume by current standards, yet they provided major challenges in the analysis which had to be done by hand using limited mathematical techniques which were then available. Johannes Kepler used Tycho Brahe's data, which could be accommodated in a few notebooks to derive his laws of planetary motion.

The amount of data increased sharply with the use of photographic plates in recording observations, as surveys could then be made of much fainter objects. The next stage of development was the use of electronic instruments for the detection. These included CCD cameras, which are much easier to use than photographic plates, and have greater stability and efficiency in light collection, so that fainter objects can be observed with a given telescope. The data from instruments like CCD cameras is obtained in digital form and therefore be easily stored and analysed. In the recent years, very large area CCD detectors which are mosaics of smaller CCDs have become available, so that surveys of the sky can be conveniently carried out. These developments have led to vast increase in data at optical and near-infrared wavelengths, where CCDs can be used. Similar technical developments have led to high data collection rates in other parts of the electromagnetic spectrum as well. We will now briefly summarise some important astronomical surveys and projects which have been generating vast volumes of data, or are expected to do so when the facilities which are being built at the present time become ready for observations over a decade. Our aim will be to indicate the kind, type and quantity of data generated without providing astronomical or technical details.

Sloan Digital Sky Survey (SDSS, www.sdss.org): This is a survey of large areas of the sky being carried out in the Northern Hemisphere with a large format Mosaic CCD with a 2.5 m telescope in the USA. The survey has produced images in five bands of the optical regions of the spectrum. The 13th release of the survey, which contains data obtained until July 2015, covers more than one-third of the celestial sphere. It provides imaging data on more than 469 million unique, primary objects including stars, galaxies, quasars, etc., and more than 4.2 million spectra. The survey has produced about 125 terabytes (TB) of data which are freely available online for any interested user, and thousands of research publications have emerged from the use of the data over the last 20 years.

The Catalina Real Time Synoptic Survey (CRTS, crts.caltech.edu: The Catalina Real Time Survey (CRTS), which is a repeated survey of the night sky conducted by a collaboration led by Caltech, aims to search for near earth asteroids and as a by-product detects a large number of transient astronomical sources. The image repository of CRTS is open to the public and is served from IUCAA. The CRTS data holding at IUCAA is currently over 30 terabytes and is expected to grow several folds as the survey continues to operate in the coming years.

The Large Synoptic Sky Survey (LSST, https://www.lsst.org): This survey will be carried with a 8.4 optical telescope which is presently being built. It will repeatedly observe areas of the southern sky which will cover 20,000 sq. degrees, with each part visited about a 1000 times in 10 years. The survey will generate a staggering 15 TB of data per night, producing about 200 petabytes (PB) of usable data over the period of the survey.

The Giant Metrewave Radio Telescope (GMRT, www.gmrt.ncra.tifr.res.in): This radio telescope was built and is operated by the National Centre for Radio Astrophysics (NCRA) in Pune, which is a centre of the Tata Institute of Fundamental Research, Mumbai. It consists of an array of 30 parabolic dishes spread over about 25 km in a specific formation. It generates about 4 TB of radio images and spectra and about 40 TB of timing data on radio pulsars. The GMRT has recently been upgraded and is expected to now produce data at about 10 times the earlier rate.

The Square Kilometre Array (SKA, skatelescope.org): This immensely large radio telescope will consist of about 3000 dishes, each 15 m in diameter, distributed over parts of South Africa and Australia. It will have capabilities far greater than any existing radio telescope and will be partially ready for observations beginning about 2020. The raw data generated by all the dishes will amount to about 15 TB per second, and the archived scientific data will be about 500 PB per year. The data volumes generated by the SKA will be far greater than data generated by exisitng surveys, and will require hexaflop computing power for the processing.

ASTROSAT (astrosat.iucaa.in): This is the first Indian satellite fully dedicated to astronomy. Launched by the Indian Space Research Organisation (ISRO) in September 2015, it has five instruments on board which produce data in the ultraviolet and X-ray parts of the electromagnetic spectrum. Four of the instruments are co-aligned, so that they can simultaneously observe a given astronomical source. ASTROSAT will produce about 200 TB/year, and more than 1 PB of scientific data will be archived over the mission lifetime.

Gravitational Wave Data (https://www.ligo.caltech.edu): Gravitational waves are ripples in the fabric of space-time and are fundamentally different from electromagnetic waves. Gravitational waves are detected using laser interferometric detectors, like the LIGO detectors in the USA and VIRGO in Italy. The LIGO-India project will set up a LIGO detector in India, which will work in tandem with those in the USA. Each detector produces data at the rate of about 10 MB per second which aggregates to about 1 PB per year of scientific data for the two LIGO detectors in operation at present in the USA.

Some of the above observatories and facilities, and many more that we have not mentioned here, are continuously producing large volumes of data, and the data production rate will increase vastly when new facilities like the LSST and SKA go on stream. All these data have to be safely archived and made easily accessible to astronomers, which is done by data centres.

3 Data Centres

At the present time, there are about 50 major data centres, and many smaller ones, which archive astronomical data. Each of these data centres typically stores a few to a few hundred terabytes of data in the form of images, spectra and catalogues. They can also store raw data which have to be processed to a form which can be used in astronomical research.

A data centre can hold data which is from one or more specific telescopes belonging to a particular observatory. Or it can hold large collections of data from different telescopes and observatories. Some data centres can be thematic, holding only data from space observatories, say, or from specific parts of the electromagnetic spectrum, like the radio or infrared. The data collection is sometimes available only to authorised users while some data centres provide data to anyone who may want to use the data. Present day data centres often provide software tools for the analysis of the data they archive. A user can obtain the required data and use either standard tools provided by the data centre or other sources or tools developed by the user for analysis of the data. Currently, it is often possible to carry out the analysis online, which obviates the need for obtaining and installing the software tools which often have complex dependencies and may require specific expertise for installation and maintenance.

One of the most comprehensive data collections is provided by the Centre de Données astronomiques de Strasbourg (CDS, cdsweb.u-strasbg.fr), which the CDS hosts the *SIMBAD* astronomical database, which is a reference database for the identification of astronomical objects; the *VizieR* service for the CDS reference collection of more than 15,000 astronomical catalogues and tables published in academic journals; and the *Aladin* interactive software sky atlas for access, visualisation and analysis of astronomical images, surveys, catalogues, databases and related data. CDS provides several useful services like the X-match service, which enable the different catalogues to be searched for common objects on the basis of their positions in the sky. The NASA Space Science Data Coordinated Archive (NSSDCA, https://nssdc.gsfc.nasa.gov/) provides data from various space missions, including astrophysical missions like the Chandra X-ray Observatory, solar and space physics data and lunar and planetary science data. The NSSDCA also provides links to centres for data analysis like the High Energy Astrophysics Science Archive Research Center. The Canadian Astronomy Data Centre (CADC, http://www.cadc-ccda.hia-iha.nrc-cnrc.gc.ca/) provides access to data obtained from many observatories and telescopes and advanced data products. It specialises in data mining, data processing, data distribution and data transfer of very large astronomical data sets. There are many astronomical data centres providing different kinds of data and it can be challenging to find the required data and services for a specific investigation, as these can be distributed over many data centres. Virtual Observatory tools provide easy ways to make these searches.

4 Virtual Observatories

Over recent years, astronomers have developed, through the Virtual Observatory (VO), standards and tools for the use of large volumes of data. These include schemas for objectively labelling the data facilitating access to it, and protocols for efficient transfer over the Internet; registries which maintain information about the location and content of various databases; and tools for statistical analysis and visualisation, subdomain-specific software packages for scientific investigations and portals for ease of access. All the developments have been done keeping in mind the use of computers for much of the data-related work, which requires all descriptors to be machine readable. The emphasis has been on interoperability, which enables astronomers to easily combine information from different domains such as radio, optical and X-ray astronomy, which is essential for multiwavelength understanding of astronomical objects.

It is now possible for an astronomer using VO registries to search for required data to obtain its location, download the data using the Internet and use VO and other tools to analyse the data. Data centres often allow users to carry out some basic analysis on their sites, so that the transfer of large volumes of data over the Internet can be avoided. A number of data visualisation and statistical analysis tools have been developed, so that users can easily extract the maximum information from their data, without having to develop the software themselves.

VO standards and services are also important for data centres, as they provide an interoperable framework for the exposure of their data. A data centre can store data using a database management system of their choice. But by using a VO wrapper on the data base, it can be made easily locatable through registries and accessible to users who would again be using VO compatible systems. A result is that finding, accessing and using data become straight forward and standard, thus ensuring complete interoperability.

The VO standards and tools have all been developed by VO projects in different centres in several countries in the world. The projects work under the umbrella of the International Virtual Observatory Alliance (IVOA, ivoa.net). The IVOA has a number of working groups and interest groups, which have members from the VO projects. This arrangement ensures that the developments made by the different projects adhere to common IVOA standards and are complementary. Over the years, a large number of standards, services and tools have become available to astronomers through the efforts of the IVOA and its member VO projects.

5   Big Data Analytics

The processing of astronomical images and spectra observed in different wavelength ranges, like optical, radio or X-ray, presents great challenges, with completely different techniques involved in the preprocessing of the data, depending on the type of telescope and instruments used in obtaining the images. After the preprocessing, photometric and spectroscopic techniques are used to obtain parameters which describe various physical properties of the objects, their chemical composition, the properties of any gas that may be present and so forth. These parameters are then subjected to statistical analysis to find patterns and are compared with theoretical predictions and the results of computer simulations.

The modern astronomer has access to various services and tools, including those which are VO compatible, to aid in the process of analysis. These tools have mainly been developed over the last few decades, during which astronomical data has increased from kilobyte sizes to gigabytes. As we have seen above, data volumes are increasing at an ever-accelerating rate and will soon routinely be in the petabyte range. These are big data volumes, and their processing requires fundamental changes in the way the data is stored, accessed and analysed. Conventional techniques are too slow to cope with such large data volumes. It is necessary to use a framework like Hadoop for the distributed storage and processing of the data. The visualisation of the data is another complex matter, with the multidimensional data cubes used to store the data, which can then be visualised using different projections. On the analytical side, sophisticated statistical techniques are needed to spot trends and relationships in multidimensions. It is also important to spot outliers which do not follow the usual correlations. These outliers, which are sparse in number, have to be spotted amongst billions of objects, but once found they would be the most interesting to study, since they would point to new classes of objects and phenomena.

The large data volumes require automation of routine astronomical tasks like image classification, for example, to distinguish between stars and galaxies. A galaxy is a very large object containing hundreds of billions of stars, but when it is very far away it can be almost unresolved and appear to be very similar to a single star. Expert astronomers can with some difficulty distinguish between the two, but when a large number of objects have to be classified, it is necessary to use, for example, an artificial neural network trained to distinguish between the two kinds of objects. Such machine learning tools can be used in various tasks of classification and regression. Ever more efficient and fast tools are required as data volumes grow, and there is the need for very quick classification, for example, in the study of transient sources. Deep learning tools, which can work directly on images and other data without the need for parameterisation, are growing in importance in the analysis of big data.

6 Big Data in Other Domains

It is estimated that about half the world's population, that is about 3.7 billion people, use the Internet. While some of this use is for scientific, technical and academic purposes, much of the use comes from social media. It is estimated that YouTube users upload 400 h of new video content each minute, Instagram users like 2.5 million posts every minute, nearly 6 billion Facebook posts are liked each day, around 4 million Google searches are conducted worldwide each minute, and about 4 million text messages are sent each minute in the USA. All this adds up to about 2.5 exabytes of data per day (1 exabyte = 1000 petabytes = a million terabytes = a billion gigabytes). Much of the data is unstructured, meaning that the content of a message or other unit of data can be of many different kinds, including text, images, video and audio. Such data are much more difficult to deal with than the rather uniform data that astronomers are used to. Social media data contain very important information about human behaviour, relationships, needs and aspirations, and the tools and expertise required to decipher the information can be complex, and sometimes different from the tools used by astronomers, though there is much that is common in the management and treatment of the data.

A scientific domain in which big data is generated is biology. It took 13 years of work in 23 laboratories, at a cost of about US $ 3 billion, to sequence a human genome for the first time. Now, human genome sequencing can be done routinely with a single machine in about 6 min and generates about 3.5 gigabytes of data. It has been estimated by Z.D. Stephens et al. (PLoS Biol. article id 10.1371, 2015) that by 2025 between 100 million and 2 billion human genomes could have been sequenced, amount to 2–40 exabytes of data. There are many other fields of biology, biophysics, crystallography, atmospheric and earth sciences, particle physics and other areas which generate equally large amounts of data. These volumes are comparable to the largest volumes of astronomical data currently expected. Given the nature of the data, it should be possible to adapt astronomical data management techniques and the VO experiences, services and tools to the large databases from other domains. Some attempts at such unification are already in progress, and there are forums like the Research Data Alliance (https://www.rd-alliance.org/) for bringing together data producers and users from different domains.

7 Further Reading

This is fast changing field with ever-growing literature, so it is difficult to point to a few comprehensive review articles which cover the field well. We have therefore provided links to useful sites in various relevant places in this article which should be able to lead the readers in the directions they may want to go.

Acknowledgements I wish to acknowledge long collaborations with Ashsish Mahabal, Yogesh Wadadekar, Nina Sajeeth Philip, Sudhanshu Barway and Kaustubh Vaghmare. T. M. Vijayraman, Anand Deshpande and others which have helped me develop some knowledge in this field.

Preserving for a More Just Future: Tactics of Activist Data Archiving

Morgan Currie, Joan Donovan and Brittany Paris

1 Introduction

The ownership of the life cycle of data—who creates it, processes it, visualizes it, preserves it—is a question of enormous power in our data-saturated society. The task of collecting, analyzing, and safeguarding large sets of data typically falls within the purview of data professionals: data scientists or statisticians tasked with deciphering the insights of large numbers or variables. Acting like modern-day oracles, they must know what metadata to collect, anticipate the needs of future users, and carefully track changes as data is worked upon and contexts change.

In parallel with the growing professionalization of data analysis, grassroots publics are also increasing the ranks of who collects, processes, and visualizes data. This work involves citizens operating outside of large institutions of government, universities, and commercial enterprises to generate and manage numerical and statistical data about matters of concern. Because numerical data is perceived as both objective and authoritative, these statistical accounts give civic groups a way to speak persuasively to governments and broader publics. Their activity, variously called data activism, civic data, or statactivism, calls attention to the political dimensions of data: the important questions of *who* or *what* gets represented *how*. Academic research has begun to document examples of data activism. In this body of the literature, most of the activity revolves around civic data collection and analysis: volunteer surveyors gathering information on the socioeconomic damages after Hurricane Sandy [15]; grassroots efforts to capture more comprehensive statistics on those killed and injured by police [5]; citizens who use industry-grade devices to detect signs of radioactivity in Fukushima, Japan [11]. Within the related realm of civic technoscience, citizens have been documenting the health effects of the chemical, oil, and gas industries within their neighborhoods [7].

M. Currie (✉) · J. Donovan · B. Paris
University of California, Los Angeles, CA 90095, USA
e-mail: msmorgancurrie@ucla.edu

© Springer Nature Singapore Pte Ltd. 2018
U. M. Munshi and N. Verma (eds.), *Data Science Landscape*,
Studies in Big Data 38, https://doi.org/10.1007/978-981-10-7515-5_5

In this scholarly literature, the activist work involves the *generation* of data to create new statistical representations or to challenge official ones. Much less attention has been focused on activity at the other end of the data life cycle spectrum: data *archiving* and preservation as an act of political critique. In fact, the literature on data activism has widely ignored issues of stewardship and long-term preservation, which raises the important question of whether largely ad hoc, democratic groups of data activists can manage data with sufficient integrity and sustainability over the long term.

This article analyzes grassroots activism committed to data archiving. These projects are devoted to the long-term custody of a dataset or multiple datasets; they may explicitly deploy archival principles to safeguard data, and in some cases, they build software designed with the logic of archival principles and practices—such as chain of custody, provenance, checksums, and multiple copies. We find that activist data archiving can create path-breaking archival models while leveraging the collective capacity for distributed networks to crowdsource this work.

To illustrate, we focus on two groups: Fatal Encounters and the Environmental Data and Governance Initiative, or EDGI. Fatal Encounters is a semi-structured organization of volunteers that maintain a comprehensive database on police killings in the USA. The database demonstrates the magnitude of the problem of missing data and critiques institutional policies that intentionally withhold data from the public. Our second example, EDGI, is an almost entirely volunteer, international network that investigates potential threats to the scientific research infrastructures needed to design and enforce environmental and energy policy. EDGI emerged in response to fears that the federal government might take action to remove data from public view in order to foreclose on scientific futures. The initiative has designed archival protocols and software to overpower the threat of disappearing federal scientific data.

We begin by defining *activist data archiving* within the literature on civic science, critical data studies, and archival theory. This body of existing research helps us make a few insights: Activist data archives persist independent of traditional institutional contexts; these non-expert publics conduct statistical work considered of high enough quality to make legitimate claims; and this work often takes shape using decentralized platforms and governance structures. To illustrate these points further, in later sections we describe the work of Fatal Encounters and EDGI and focus on the archival protocols and workflows their members developed.

2 Data Archiving for Radicals

Activist data archives, admittedly, are chimeras; they are hybrid beasts that mutate as new elements are introduced, and as such their definition is hard to pin down. Where do we draw the difference between a digital spreadsheet persisting on a public server and a data archive? Is an activist who maintains a database of social justice-related statistics also an archivist? We do not seek out essentialist definitions here. A *data archive* can be defined simply as the long-term storage of qualitative and quantitative datasets. To define an *activist data archive*, in particular, we draw on the existing literature in critical data studies and archival theory before looking at current examples of these activist data archives in action.

In media and critical data studies, scholars have formulated new terms to describe data practices that ignite activism and resistance, variously labeling this political work counter-data [12], statactivism [3], or data activism [16]. The rise of data activism has come about as statistical devices such as GIS, online survey tools, and other platforms for crowdsourced data collection are now relatively affordable and widely available. Data activism encourages communities to conduct their own research on matters of concern and to build alternative statistical narratives than those offered by mainstream science or industry. Data collection and analysis is one way that lay publics can have a voice in scientific debates and public policy. In the process, data is revealed as a political tool that shapes reality as much as it reflects consensus around our fundamental assumptions about it. Activist data archiving can be analyzed as a particular form of data activism.

Activist data archiving can also be understood as a form of archival activism. There are clear affinities between radical archival practices and data activism as they focus on including interested stakeholders and allowing them the informational literacy needed to assert their points of view. Radical engagements with archival practices coalesce around a notion of activism that informs archival description and representation, collection, and curation. Archival activism often entails the collection and preservation of alternate representations of a group, events, or issues. These radical archival practices are necessarily political and often intentionally unaffiliated with institutional support. Thinking through these historic practices can help us refine our understanding of activist data archiving.

From this literature, we have drawn out three general dimensions of activist data archives that distinguish them from other types of data archives. Later, we offer our two examples, Fatal Encounters and EDGI, to illustrate and refine these points.

2.1 Activist Data Archives Persist at the Grassroots Level

Like counter-data and statactivism projects, activist data archives typically fall somewhere within the civic realm, a space where people organize voluntarily outside of government and the private sector. This sphere encompasses nonprofit organizations, global NGOs, political organizations, philanthropies, international standards bodies, and grassroots political activism. The archival projects we describe in this article persist at the grassroots level, where most of the work is conducted by organized publics [9] who commune voluntarily around shared matters of concern, with little-to-no paid staff or formal bureaucracy.

There is a political rationale for the importance of autonomous, grassroots archival activity. In archival theory, *community archives* function as grassroots repositories that allow certain groups—particularly the LBGTQ population, indigenous peoples, or ethnic minorities—to take control of their historical representations. Community archives often form in response to cultural silencing and marginalization; their participants have a palpable connectedness to a shared history and cultural legacy. Community archives often exist independent of government or scholarly institutions; they persist, rather, based on the desire to create a collective identity and challenge the dominant narratives of the past. One well-known community archive is the Mazer Lesbian Archive, which accumulated in a residence in the Altadena neighborhood of Los Angeles throughout the 1980s by volunteers who documented largely invisible lesbian culture (Today, the Mazer is held by the University of California, Los Angeles, in a case of an archive shifting from the hands of an autonomous, grassroots community to a formal institution.)

Archival theorists also use the term "activist archiving" to describe archives that are necessarily political, progressive, and maintained by and for the very people who generate them [19]. Activist archives are often intentionally unaffiliated with institutional support. Autonomy is a key to the success of community and activist archives. Past and present marginalization, colonization, slavery, and violence to particular minority communities remain central to institutions of American democracy—including universities or federally funded historical archives. For this reason, institutions cannot always be counted on to meaningfully memorialize on behalf of these voices. By remaining independent from formal institutions, archivists are making a statement about how entrenched organizations play a role in their political necessity in the first place. Drawing from these radical archival practices, archival data activism can also encompass the long-term preservation of data that addresses social justice issues while working outside of formal organizational structures.

2.2 Activist Data Archives Bridge Lay Publics and Experts

For any grassroots archival project, the matter of legitimacy and voice arises: How do activist archives, operating outside of institutional norms, prove their trustworthiness as legitimate sources of information that can influence policy or broad public opinion? How are sound protocols and standards established by a base of volunteers? In order to take a productive role in political controversies, grassroots data archives must present their information as legitimate enough to appeal to officials and wider publics. The issue will not only define the operational structures of the grassroots data archive, it also determines how much political power it wields.

The literature on data activism and civic science has widely addressed these issues of legitimacy and participation in the public sphere. First, these scholars point out that data itself has a rhetorical force that can translate concerns of affected publics so that they are understood by experts and wider society; this is because statistical data bestows a measure of objectivity by granting the author the appearance of a distanced perspective. According to Daston and Galison [6], "mechanical objectivity" removes the subjective interpretation from phenomena, thus permitting a machine or statistical patterns to produce "better observations" than humans (ibid., p. 83).

However, data is more often the output of what Latour [14] calls "centers of calculation," powerful spaces where inscriptions are collected, sorted, analyzed, and made meaningful (p. 32). Centers of calculation, such as government administrations, universities, think tanks, and institutes, bring together different networks of people, concepts, and objects that are distributed across time and space in order to create and circulate knowledge. Government agencies, for instance, often call upon objectivity through statistics in order to quell public criticism, substantiate current practices, and distribute resources.

Journalists, activists, and critics looking to generate arguments that appeal to the state's centers of calculation—or to challenge authority and the status quo—must mobilize their own statistics [1]. Data activism invokes Certeau's concept of tactics: weapons deployed by those traditionally lacking in power or without the financial or social capacity to form long-term strategies [4]. Statistical claims in this way expand the voices of those who contribute to authoritative metrics on a phenomenon; they can redress the imbalance between government or scientific claims and those produced by grassroots efforts.

Data activists must develop rigorous methodologies, create their own centers of calculation, and provide ample metadata to make claims to objectivity that are taken seriously by those with the power to make laws and policy and change institutions. Civic technoscience projects serve as a model here: They deliberately involve non-experts and community-based groups in scientific questioning and data production, using devices to gather data that is robust and scientifically vetted [22]. "Bucket brigades," as one example, equip citizens with EPA-approved buckets to collect air samples to measure for toxic emissions. The brigades emerged out of a practical necessity to confront the lack of quality tools that could allow non-scientists—especially citizens affected by pollution—to participate in air monitoring. The Louisiana Bucket Brigade, organized around the 2010 Deepwater Horizon oil spill, collectively interpreted and contextualized bucket data to build a strong case against industry that harm was being done to their health and local environment [18].

The literature on data activism is limited because it does not help us explain data *archiving* as an activist project. For grassroots data archives, legitimacy goes beyond how the data is collected and analyzed. To be trustworthy, the archivist will want to show that a dataset has not been tampered with or degraded in any way; she will want to use professional standards for digital data preservation: metadata that documents chain of custody and provenance; metadata that preserves context; analog and automated checksums; and multiple, mirrored copies to prevent loss. Grassroots data archives may even create their own software to fulfill these archival practices. In the case of EDGI described below, for instance, volunteers designed bespoke platforms for data ingestion and documentation in order to ensure the integrity of the data.

2.3 Activist Data Archives are Decentralized

The work of activist data archiving creates new models for digital archiving by leveraging the logics of distributed networks. The networking capacity of the Internet has opened, if not new, at least expedited, opportunities for groups of like-minded people to come together and work on projects collaboratively. As a result, the Internet provides tools not only useful to centers for calculation, but also to "centers of coordination," where technologies reorient space and time so that projects can be done within a stable environment, but from a distance [20].

In contrast to the tools, methods, and sources available to journalists and activists in the 1990s, contemporary activists have access to networked communication technologies and a wealth of digital ephemera that allow for data archiving as a distributed project of people working in any locale. Technologies of social change —tools and knowledge employed to bring about justice—lay at the heart of projects that rely on the connective capacity of the Internet to link together people, information, and resources in the service of the common good. For this work, which relies primarily on participants who go financially uncompensated, the more people who are willing to do the work, the more robust and sophisticated the archive can be.

In some cases, archival activist work relies on nonrepresentational governance structures by participants so that data standards and workflows are ad hoc creations that result from participant deliberation. Distributed, though, does not mean disorganized. These networks will have a person or set of people who take on important tasks such as maintaining and overseeing the server where the data resides, taking on administrative tasks involving labor coordination, or software design.

The technical platforms used often reflect the organizing logic of decentralized participation. Cloud software such as Slack supports group communication; Google spreadsheets allow participants to work collectively on large CSV files that streamline the process. The real innovation here is that the spreadsheet is "spreadable media" itself; it is circulated across media platforms open to inspection and criticism by variety of contributors and viewers [10]. In other examples, participants might devise workflows and databases using open-source software they design; in the case of EDGI, described below, participants designed the software to take archival principles of chain of custody and provenance into account.

Decentralization is a basic record management principle: have multiple copies of the data in multiple places. Often the archive uses open licenses and encourages mirroring so that it can exist on a multitude of servers, not attached to any one central institution. Archival metadata ensures that copies remain updated, citable, and have a chain of custody.

In the next sections, we describe two projects that further illustrate these dimensions of activist data archives.

3   Fatal Encounters

Statistics are useful to those who seek to counter powerful state narratives regarding the prevalence of police killings and the threats posed to police by citizens. Not only are arrest-related deaths (which activists routinely call "police killings") purposefully not aggregated by the federal government, but also raw numbers are difficult to obtain even when required by law. While arrest-related death is arguably a normal consequence of some emergencies, these numbers seem astoundingly large because there are no national comparisons across years. Instead when compared to other countries, the American situation is bleak; only 55 people died in police custody in England and Wales over the past 24 years [13]. Among the groups counting American police killings in 2015, the number of deaths ranges from 990 to 1356 for a single year.

Because complete data on police brutality and police killings remains out of reach to journalists and activists, they must devise methods for locating and recording these events outside the jurisdiction of police. The Website *Fatal Encounters* took up this challenge by collecting, sorting, and analyzing data on police killings and providing this information in publicly accessible databases online.

Fatal Encounters is a comprehensive clearinghouse of information begun in 2000. Participants continuously update the database, which operates as a living archive. A small number of volunteers, supported mainly by public donations, compile and compare open sources online to locate the missing masses. Fatal Encounters utilizes shared documents, particularly large CSV files, as well as editing protocols, and agreements about workflows in order to maintain the archive.

To collect the data, Fatal Encounters' volunteers draw from a multitude of sources: Freedom of Information Act requests, public records, police records, media reports, coroner reports, social media submissions, photographs, original reporting, and crowdsourced verification. Fatal Encounters' founder double-checks all the reports received against local news stories before publication. As a result of limited resources and methodological rigor and complexity, the Fatal Encounters' database lags about three months behind other, similar counts (such as the one maintained by *The Guardian* and *The Washington Post*). Fatal Encounters, however, tends to track and record more data than other groups. Because Fatal Encounters values exhaustive and comprehensive verification, some cases may stay in the database until such time that the cause of death has been clarified, either through public records, FOIA requests, or updated media reports. This complementary method of collection produces data redundancy that further legitimizes the reliability of their database.

Fatal Encounters fits our criteria of an activist data archive. The Website is the result of grassroots labor by participants who understand its rhetorical force. The project was volunteer run and paid for by its founder journalist Brian Burghardt from its inception until 2014, when it started to receive grant and crowdfunding. Burghardt is paid to oversee the project, but the brunt of the work still relies on volunteers. The work is nevertheless robust, and it has become a go-to source on police-officer-involved homicides for the news media. Finally, the work is decentralized, both in terms of its labor pool—contributors live around the world—and in terms of the material database: volunteers submit to a form that on the organization's Website that populates a CSV file that streamlines the metadata. The dataset is backed up and maintained by Burghardt and a technical volunteer.

4   EDGI

The Environmental Data and Governance Initiative (EDGI) is a collaborative network that investigates potential threats to the scientific research infrastructure necessary for environmental and energy policy in the USA. EDGI began in November 2016, soon after Donald Trump was elected to the US presidency. Some of EDGI's founders hail from Canada, where they have fresh memories of former Prime Minister Stephen Harper's climate change skepticism. Harper's policies led to the physical destruction of materials from scientific libraries and archives and silenced many government climate scientists. Internationally, scholars shared similar concerns that Trump's ideological position on climate change would result in the removal of this information from public access. EDGI formed with a few goals in mind: to design Web-based tools, foster a network of researchers, and host public events in an effort to keep public environmental data accessible.[1]

EDGI members devote themselves to several tasks: Website monitoring for any changes to content on a subset of federal Websites; archiving datasets and Web pages alongside local organizers and volunteers at archiving events, called DataRescues, and in collaboration with other Internet archiving efforts, including Internet Archive, the University of Pennsylvania's DataRefuge, and Climate Mirror; following and documenting policy around federal data governance; and interviewing federal experts. This discussion will focus on EDGI's data archiving activity in particular.

To be clear, federal data is not so much in danger of being deleted than of becoming difficult to access thanks to policy changes and related funding cuts. EDGI's concern is that data that previously circulated as "open" will disappear from the Web in an effort to reduce the capacity for scientific research and calls for reform and regulation. Access to information—though not necessarily the original data itself—can also deteriorate through link rot, or broken Web page links. Once data goes offline, it might still be available through FOIA requests, but FOIA requests can be denied, and requested items can be redacted.

---

[1]See EDGI's Report, "Introducing the Environmental Data and Governance Initiative," published on February 1, 2017. https://envirodatagov.org/publication/introducing-edgi/. Archiving was only one part of EDGI's initial goals.

At the time of this writing, one dataset has gone down: the USDA's animal welfare reports [2]. Multiple EPA climate change subdomains have been removed and now redirect to an "update" page [17]; as a result "A Student's Guide to Global Climate Change," an educational site for kids with more than 50 Web pages, is currently not accessible from the EPA's Website [8]. Links to the State Department's "Climate Change" pages also stopped working after the transition to the Trump administration's whitehouse.gov [21]. It should be emphasized that the EPA, White House, and State Department Websites all contain archives of Obama-era Websites, so they are still accessible and not deleted. However, these changes make this information harder to access in the public domain.

One vital part of EDGI's archiving work involved the coordination of "Datarescues" around the USA; over 30 of these occurred from December 2016 until June 2017. Often hosted by higher educational institutions, DataRescues invited volunteer participants together to copy federal scientific datasets, documents, and Web pages using an array of archiving tactics that EDGI participants call "seeding," "harvesting," and "bagging." "Seeding" is a term EDGI members use for nominating federal Web page URLs to the Internet Archive's End of Term project. End of Term routinely archives.gov Websites during periods of presidential transitions. The project uses an automated Web crawler to "crawl" or map out relations between federal Web pages, as well as archive Web snapshots; it also accepts nominations for URLs from external parties who want to ensure particular Web pages are replicated. EDGI volunteers contributed to the End of Term project by nominating the pages of agencies devoted to energy and environmental research in particular. Seeding at DataRescues entailed an enormous amount of coordination. EDGI volunteers devised "primers" that mapped agency Websites in order to break up the hundreds of thousands of federal Web pages into meaningful chunks; they also created a coded organization chart to draw relations between different agencies and agency structures in the .gov domain.

The Internet Archive's basic crawler, however, encounters many intractable Web pages that it cannot crawl. During the seeding process, participants "nominate" Web pages containing content that cannot be archived by the Internet Archive. Participants then must "harvest," or download, the data themselves. In order to archive this uncrawlable data, EDGI participants designed an open-source Web application called Archivers.space. Archivers.space is a project management tool that tracks a dataset's process from its uploading to a server through multiple stages of research and vetting by participants. In one step, participants "bag" the data—EDGI's term for running a python script that generates checksums. Another step entails pulling the data into a zip file along with information that pertains to its Web context, metadata, and any other important descriptive information. In the final step, the dataset is fully described, including its chain of custody, context, and provenance (in future iterations, EDGI plans to automate parts of this step). Once a dataset is ready, it is pushed to DataRefuge.org (maintained by DataRefuge), where scientists and activists can use it.

EDGI also fits our definition of an activist data archive. The archival work and software design rely entirely on volunteers, including people with backgrounds in environmental justice, radical activism, and open-source cultures who share a commitment to horizontal organizing as a core principal. EDGI's governance form is also responsive to the logic of networked digital data; members use Slack to communicate and come to consensus on many archival processes. This decentralized form allows participants—some of whom have experience in data science, digital curation, librarianship, and government records, while others have no technical background at all—to collectively weigh in on some of the archival protocols that have developed.[2] The data archive reflects EDGI's decentralized form; it can be mirrored on any server so that it is not attached to a single, central institution. The Archivers.space software is open source with open licenses, so anyone could, theoretically, duplicate these efforts.

EDGI is also different from Fatal Encounters, in some interesting ways. Fatal Encounters participants both produce and maintain one dataset, while EDGI's participants are only involved in archiving datasets that they had no hand in creating. Fatal Encounters sets out to augment and correct any account of police killings by the government. Fatal Encounters disturbs, not reinforces, the federal government's own calculations. EDGI, on the other hand, archives data created by US-funded scientists to document evidence of climate change and human-induced ecological violence; it replicates and preserves scientific evidence to protect it for future use, not to augment, challenge, or reinterpret it. EDGI's Web archiving and data mirroring pluralizes and distributes the material context of the data; its political work lies in decentralizing information by removing it from institutional control, not re-presenting it.

5   Conclusion

Archival data activism alters the conditions of possibility for political agency in an increasingly data-driven society. While data holds immense power when applied from the top-down, the above examples show that there are ways that this power can be wielded from the bottom-up; it is preservation work that promotes coalition building, grassroots data literacy, and community-based organizing in situations where communities of interest are geographically dispersed.

Both of the examples described above meet power with power, that is, answer entrenched and often unjust political power with carefully mobilized and curated data that enables arguments for social justice. Activist data archiving provides methods that can unite people to engage in common causes—harkening to new ways to envision how "The people united, can never be defeated," a line from the

---

[2]Much of the archival protocol work was also stewarded by DataRefuge. EDGI's consensus processes were not used in those instances.

old protest song that speaks of democratic movements from below and has been used recently in both protests against climate change and police brutality. While the possibility for defeat is a real one, collectivizing around data is one way for interest groups to build power in the fight for a more just future.

References

1. Baudot, P.-Y. (2014). Who's counting? Institutional autonomy and the production of activity data for disability policy in France (2006–2014). *Partecipazione E Conflitto, 7*(2), 294–313. https://doi.org/10.1285/i20356609v7i2p294.
2. Brulliard, K. (2017, February 9). USDA removed animal welfare reports from its site. A showhorse lawsuit may be why. *Washington Post.* Retrieved from https://www.washingtonpost.com/news/animalia/wp/2017/02/09/usda-animal-welfare-records-purge-may-have-been-triggered-by-horse-industry-lawsuit/.
3. Bruno, I., Didier, E., & Vitale, T. (2014). Statactivism: Forms of action between disclosure and affirmation. *In Partecipazione e Conflitto. The Open Journal of Sociopolitical Studies, 7*(2), 198–220.
4. Certeau, M. de. (1984). *The practice of everyday life.* California: University of California Press.
5. Currie, M., Paris, B. S., Pasquetto, I., & Pierre, J. (2016). The conundrum of police officer-involved homicides: Counter-data in Los Angeles County. *Big Data & Society, 3*(2), 2053951716663566. https://doi.org/10.1177/2053951716663566.
6. Daston, L., & Galison, P. (2010). *Objectivity.* Zone Books.
7. Dosemagen, S., Warren, J., & Wylie, S. (2011). Grassroots mapping: Creating a participatory map-making process centered on discourse. *The Journal of Aesthetics & Protest,* (8). Retrieved from http://joaap.org/issue8/GrassrootsMapping.htm.
8. Eilperin, J. (2017, May 6). The EPA just buried its climate change website for kids. Retrieved June 24, 2017, from https://www.washingtonpost.com/news/energy-environment/wp/2017/05/06/epa-buries-climate-change-site-for-kids/.
9. Emerson, S. (2017, April 27). The interior department just quietly scrubbed its climate change page. Retrieved June 24, 2017, from https://motherboard.vice.com/en_us/article/kbvkp3/the-interior-department-scrubbed-its-climate-change-page-doi-zinke.
10. Jenkins, H., Ford, S. & Green, J. (2013). *Spreadable media: Creating value and meaning in a networked culture.* US: New York University Press.
11. Kawano, Y., Shepard, D., Shobugawa, Y., Goto, J., Suzuki, T., Amaya, Y., … Naito, M. (2012). A map for the future: Measuring radiation levels in Fukushima, Japan. In *2012 IEEE Global Humanitarian Technology Conference* (pp. 53–58). https://doi.org/10.1109/GHTC.2012.18.
12. Kitchin, R., & Lauriault, T. P. (2014). *Towards critical data studies: Charting and unpacking data assemblages and their work* (SSRN Scholarly Paper No. ID 2474112). Rochester, NY: Social Science Research Network. Retrieved from https://papers.ssrn.com/abstract=2474112.
13. Lartey, J. (2015, June 9). By the numbers: US police kill more in days than other countries do in years. *The Guardian.* Retrieved from https://www.theguardian.com/us-news/2015/jun/09/the-counted-police-killings-us-vs-other-countries.
14. Latour, B. (1986). Visualization and cognition: Thinking with eyes and hands. *Knowledge and Society: Studies in the Sociology of Culture Past and Present, 6,* 1–40.
15. Liboiron, M. (2015). Disaster data, data activism: Grassroots responses to representations of Superstorm Sandy. In J. Leyda & D. Negra (Eds.), *Extreme weather and global media* (1st ed.). New York: Routledge.

16. Milan, S. (2016). *Data activism as the new frontier of media activism* (SSRN Scholarly Paper No. ID 2882030). Rochester, NY: Social Science Research Network. Retrieved from https://papers.ssrn.com/abstract=2882030.
17. Mooney, C., & Eilperin, J. (2017, April 29). EPA website removes climate science site from public view after two decades. *Washington Post*. Retrieved from https://www.washingtonpost.com/news/energy-environment/wp/2017/04/28/epa-website-removes-climate-science-site-from-public-view-after-two-decades/.
18. Ottinger, G. (2010). Buckets of resistance: Standards and the effectiveness of citizen science. *Science, Technology and Human Values, 35*(2), 244–270. https://doi.org/10.1177/0162243909337121.
19. Sellie, A., Goldstein, J., Fair, M., & Hoyer, J. (2015). Interference archive: A free space for social movement culture. *Archival Science, 15*(4), 453–472. https://doi.org/10.1007/s10502-015-9245-5.
20. Suchman, L. (2011). Practice and its overflows: Reflections on order and mess. *TECNOSCIENZA: Italian Journal of Science & Technology Studies, 2*(1), 21–30.
21. Varinsky, D. (n.d.). Data on climate change progress is disappearing from the US State Department website. Retrieved June 27, 2017, from http://www.businessinsider.com/climate-action-reports-disappeared-2017-1.
22. Wylie, S. A., Jalbert, K., Dosemagen, S., & Ratto, M. (2014). Institutions for civic technoscience: How critical making is transforming environmental research. *The Information Society, 30*(2), 116–126.

Little Data from Big Data for Disaster Risk Reduction in India

Vinod K. Sharma and Ashutosh Dev Kaushik

1 Introduction

It is known to the scientific community that a major portion of South Asia is extremely vulnerable to both seismic and hydro-meteorological hazards such as floods, cyclones, droughts and derivative disasters like forest fires and landslides. This vulnerability is compounded by socio-economic conditions, which exacerbate the risk of disasters. The vulnerability varies from region to region, and a large part of the country is exposed to such natural hazards, which often turn into natural disasters causing significant disruption of socio-economic life of communities leading to loss of life and property [1].

The major disasters that affect South-Asian countries including India and Bangladesh are cyclones, floods, tornadoes, droughts and earthquakes. The CERD disaster database lists 93 disasters over the period 1986–1995. Out of these, 40 were cyclones and 31 floods. The cyclone disasters occurred in 1970 (300,000 dead) and 1991 (138,000 dead) are among the worst disasters in the world [2]. For disaster risk reduction (DRR), four priorities of actions are identified by Sendai Framework. These priorities are (a) understanding disaster risk; (b) strengthening disaster governance; (c) investing in DRR; and (d) enhancing disaster preparedness for effective response and inclusion of the concept of "Build Back Better" in disaster recovery. At every level, there is need for background data and its proper interpretation for subsequent feedback for policy with the proper data availability digital

V. K. Sharma (✉)
Disaster Management & Environment, Indian Institute of Public Administration,
I. P. Estate, Ring Road, New Delhi 110002, India
e-mail: profvinod@gmail.com

A. D. Kaushik
HY-MET Division, National Institute of Disaster Management, Ministry of Home Affairs,
Government of India, NDCC II Building, Jai Singh Road, New Delhi 110001, India

© Springer Nature Singapore Pte Ltd. 2018
U. M. Munshi and N. Verma (eds.), *Data Science Landscape*,
Studies in Big Data 38, https://doi.org/10.1007/978-981-10-7515-5_6

format, one formulation could easily generate citizen-centric information dissemination applications that are useful for diffusing chaotic situation during crisis. For instance, inclusion of electronic communication media and existing information system such as SMS, Internet/servers, phones, TV and WhatsApp are very useful in data-centric DRR. Thus, disaster management communication network can be formed in collaboration with such agencies involved in disaster preparedness at remote locations. A database resource can be overlapped on the geographically distributed database servers by setting up a sustainable development networking programme at national and regional levels. A website containing data and information of disaster management plans can assist in reducing the damages during a disaster. Information of the recovery techniques can be accumulated in the database linking with other available resources.

1.1 Disasters

Disaster, as defined in the DM Act 2005 [3], means "a catastrophe, mishap, calamity or grave occurrence in any area, arising from natural or man-made causes or by accident or negligence which results in substantial loss of life or human suffering or damage to, and destruction of, property, or damage to, or degradation of, environment and is of such a nature or magnitude as to be beyond the coping capacity of the community of the affected area".

A more modern and social knowledge of disasters, however, identifies this distinction as artificial since most disasters result from the action or inaction of people and their social and economic assets. This happens by people living in ways that degrade their environment, developing and over populating urban areas, or creating and perpetuating socio-economic aspects. Communities and population settled in such areas which are susceptible to the impact of a raging river or the violent tremors of the earthquake are placed in situations of severe vulnerability due to their socio-economic conditions. This is compounded by every aspect of nature being subjected to seasonal, annual and sudden fluctuations and also due to the unpredictability of the timing, frequency and magnitude of occurrence of the disasters.

Around 200 million young people worldwide are affected each year by reported disasters and thousands of them are died or injured. The economic costs are on the rise at an alarming proportion, which has increased by a factor of 8 at present in comparison to the 1960s. Globally, 90% of the disasters and 95% of the total disaster-related deaths occur only in the developing countries, which fall between the areas of Tropic of Cancer and the Tropic of Capricorn. This region is dominated by the poorer countries of the world, where the problems of disaster management are very solitary due to the seemingly competing needs between basic necessities for people and economic progress.

2 The Indian Scenario

The Indian subcontinent is highly vulnerable to cyclones, droughts, earthquakes and floods, etc. Avalanches, forest fires and landslides occur frequently in the Himalayan belt of northern India. Out of the 36 total States/Union Territories in the country, 25 are disaster prone. On an average, about 50 million people in the country are being affected badly by one or the other disaster every year. In the 1970s and the 1980s, droughts were the biggest killers in India, the situation, however, stands altered today. It is probably a combination of factors like better resources management and food security measures that have greatly reduced the number of deaths caused by droughts and famines. Floods, high winds and earthquakes dominate (98%) the reported injuries, with ever-increasing numbers in the last ten years. The period from 2001 to 2011 was dominated with a large number of earthquakes in Asia that have a relatively high injury-to-death ratio. However, the periods 2006–2007 and 2015–2016 were badly affected by floods, droughts, cyclones, earthquakes, landslides and avalanches and some of the other major disasters that repeatedly and increasingly affect India (Table 1).

In last few decades, there were many natural disasters caused very high mortality in India such as Cyclones of West Bengal (1970) killed 500,000 persons, Andhra Pradesh (1977) killed more than 10,000 people and Orissa Super Cyclone killed more than 10,000 people. Similarly, some of the earthquakes such as Latur (maharashtra) in 1993 and Gujarat earthquake in 2001 killed more than 10,000 people (Table 2). Besides these catastrophes, there were smaller earthquakes occurred in Uttarkashi, Chamoli and Jabalpur, and frequent floods in the north-east, Uttar Pradesh, Bihar and Kerala. Unfortunately, these disasters were not considered as learning opportunities, and hence, lessons were not learned from them in view of preparedness and in combating the future disasters. What happened in Gujarat in 2001 and the way it was handled are grim reminders of the fact that still much is left to learn and improve.

The precise cost of the disaster in terms of loss of lives, property, loss of development opportunities, etc., cannot be clearly assessed. The cost of disaster is clearly inequitable, falling heavily only on a few. Disasters result not only in loss of shelter but also create hardships, lack of food availability, temporary loss of livelihood and disrupt socio-economic activities. Some of the losses may be redeemable and compensated through disaster relief and insurance scheme. However, apart from economic dimension, such disturbances have their psycho-social dimensions as well, which need to be studied and documented besides developing appropriate mitigation strategies. There is a database of each disaster about life loss (humans/cattle) and crop damage, etc., with Ministry of Home Affairs. Likewise, similar information for droughts is available with the Ministry of Agriculture.

Table 1 Statement showing damage due to cyclone/flashfloods/floods/landslides/cloud burst, etc., during 2006–2007 to 2015–2016

S. No.	Year	No. of human lives lost	No. of cattle heads lost	No. of houses damaged	Cropped area affected (lakh hectares)
1	2006–07	2402	455,619	1,934,680	141.724
2	2007–08	3494	104,423	2,659,544	72.546
3	2008–09	3405	53,833	1,646,905	35.56
4	2009–10	1676	128,452	1,359,726	47.134
5	2010–11	2256	48,778	1,338,619	36.96
6	2011–12	1530[a]	6,976	787,290	18.85
7	2012–13	946	24,293	667,319[b]	14.44
8	2013–14	5677+	102,998	1,210,227	63.74
9	2014–15	1674	92,180	725,390	26.85
10	2015–16	1460	59,057	1,313,371	31.09
	Total	24,520	1,076,609	13,643,071	583.934

[a]This includes 60 lives lost in Sikkim, 11 lives lost in West Bengal and 10 lives lost in Bihar due to earthquake of 18.09.2011
[b]This includes 4,693 no. of huts. + This includes persons missing in the natural disasters
Source Ministry of Home Affairs (MHA)

Table 2 India's deadliest disasters (from 1900 onwards)

S. No.	Name of event	Year	State and area	Fatalities
In the known history				
1	Earthquake	1934	Bihar	6,000 deaths
2	Bhola cyclone	1970	West Bengal	500,000 deaths (including Hindu Kuch Himalayas and surrounding areas)
3	Drought	1972	Large part of the country	200 million people affected
4	Drought	1987	Haryana	300 million people affected
In the last century				
1	Earthquake	1905	Kangra, Himachal Pradesh	20,000 deaths
2	Cyclone	1977	Andhra Pradesh	10,000 deaths, hundreds of thousands homeless, 40,000 cattle deaths. Destroyed 40% of India's food grains

(continued)

Table 2 (continued)

S. No.	Name of event	Year	State and area	Fatalities
3	Latur earthquake	1993	Latur, Marthawada region of Maharashtra	7,928 people died and another 30,000 were injured
4	Orissa super cyclone	1999	Orissa	10,000 deaths
5	Gujarat earthquake	2001	Bhuj, Bachao, Anjar, Ahmadabad, and Surat in Gujarat State	25,000 deaths, 6.3 million people affected
6	Tsunami	2004	Coastline of Tamil Nadu, Kerala, Andhra Pradesh and Pondicherry, as well as the Andaman and Nicobar Islands of India	10,749 deaths, 5,640 persons missing, 2.79 million people affected, 11,827 ha of crops damaged, 3,00,000 fisher folk lost their livelihoods
7	Maharashtra floods	July 2005	Maharashtra State	1097 deaths, 167 injured, 54 missing
8	Kashmir earthquake	2005	Kashmir State	86,000 deaths (includes Kashmir and surrounding Himalayan region)
9	Kosi floods	2008	Noth Bihar	527 deaths, 19,323 livestock perished, 222,754 houses damaged, 3,329,423 person affected
10	Cyclone Nisha	2008	Tamil Nadu	204 deaths, $800 million worth damages
11	Flood	Sept. 2009	Bihar, Orissa, West Bengal	992 deaths, 1,886,000 total affected, 220 US$ million estimated damage
12	Flood	July 2010	Ambala district, Haryana	53 deaths, 400,000 total affected, 447 US$ million estimated damage
13	Flood	2011	Angul, Basalore, Bargarh	239 deaths, 3,443,989 total affected, 930 US$ million estimated damage
14	Flood	Sept. 2012	Assam, Sikkim, Arunachal Pradesh	21 deaths, 2,000,000 total affected, 98 US$ million estimated damage
15	Cyclone Nilam	Nov. 2012	Andhra Pradesh, Tamil Nadu	40 deaths, 70,000 total affected
16	Uttarakhand disaster (flood and landslide)[a]	June 2013	Uttarakhand	680 deaths, Many millions people affected and 4117 people missing

(continued)

Table 2 (continued)

S. No.	Name of event	Year	State and area	Fatalities
17	Cyclone Phailin and flood[a]	Oct. 2013	Odisha and Andhra Pradesh	45 deaths (44 in Odisha and 01 in Andhra Pradesh), 13 million people affected
18	Malin landslide	July 2014	Maharashtra	151 deaths
19	Flood	Sept. 2014	Jammu and Kashmir[c]	298 deaths and 275,000 population of 10 districts affected
20	Flood	Nov. 2015	Tamil Nadu (Chennai)[b]	422 deaths

Source Topics 2000, Natural Catastrophes-the current position, Special Millennium Issue, Munich Re Group, 1999
Centre for Research on Epidemiology of Disaster (CRED) EM-DAT, Belgium
[a]NIDM (MHA, GOI) New Delhi
[b]https://en.wikipedia.org/wiki/2015_South_Indian_floods
[c]http://ndmindia.nic.in/flood-2014/floodsOct-2014.htm

3 Vulnerability

As stated above, India is the most vulnerable to disasters in the world (Fig. 1). Disaster is the product of hazards like floods, cyclones, landslides and earthquakes and these are not rare, while the vulnerability varies from region to region. Timmenmen [4] reviewed vulnerability at society or community scale and defined it as the degree to which a system, or part of system, may react adversely to the occurrence of a hazardous event. The reciprocal of vulnerability is resilience. Vulnerability analysis results in an understanding of the level of exposure of persons and property to the various hazards identified.

3.1 India's Vulnerability

India is a country, which is highly vulnerable to natural disasters. Enormous population pressure and urbanization have forced people to live on marginal lands or in cities where they are at greater risk to damages caused by disasters. Whether a flood or a drought or a devastating earthquake, millions of Indians are being affected each moment when a disaster strikes. In addition to large-scale displacement and the loss of life, these events result in the loss of property and agricultural crops worth millions of dollars annually. As an example, the Orissa super cyclone

Vulnerability
Vulnerability of society as the sum of susceptibility, lack of coping capacities and lack of adaptive capacities

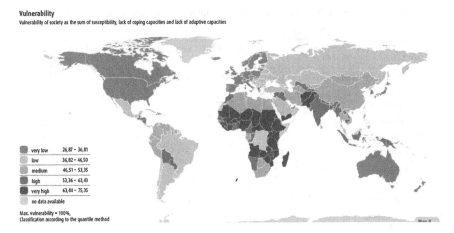

very low	26,87 – 36,81
low	36,82 – 46,50
medium	46,51 – 53,35
high	53,36 – 63,43
very high	63,44 – 75,35
no data available	

Max. vulnerability = 100%,
Classification according to the quantile method

Fig. 1 World vulnerability to disasters. *Source* Centre for Research on the Epidemiology of Disasters

in 1999 killed thousands and destroyed more than one million hectares of crops. While, Gujarat earthquake in 2001 claimed thousands of lives, left millions of people homeless and ruined public infrastructure worth hundreds of millions of dollars. These disasters typically caused the substantial loss of hard-won development gains.

India is a nation with varied climatologic and topographic conditions. Therefore, 68% of the land is drought prone, 58% is prone towards earthquake, 12% to floods and 8% to cyclones. This accounts that almost 85% of the land area in India is prone towards natural hazards while 25 States/UTs have been marked as hazard-prone areas. The major disasters in India are floods, earthquakes, droughts and cyclones while the minor disasters in India are landslides, avalanches, hailstorms and forest fires. In India, most of the states experience more than one type of disaster (Table 3).

As per the experiences of disasters seen so far in India, Bangladesh and other developing countries, it is found that when a disaster strikes, poor people are more affected. Among poor people, women are more vulnerable. In a poor family, life expectancy of men and women, literacy rate of men and women, nutritional status, mobility, access to information, etc., all go in favour of men compared to women thus making them more prone.

In this framework, disasters are observed as "the consequences of unattended risks that is neither prevented nor mitigated, and therefore, the countries and communities are not prepared". For more than a century, disaster as a subject has been researched.

Table 3 India's vulnerability: disasters occurring in different states and union territories In India

S No.	Name of state/ UT	Type of natural disaster						
		Cyclone	Landslide	Flood	Drought	Forest fire	Earthquake	Total
1	Andhra Pradesh	✓	–	✓	✓	–	–	3
2	Arunachal Pradesh	–	–	✓	–	–	✓	2
3	Assam	–	–	✓	–	–	✓	2
4	Bihar	–	–	✓	✓	–	✓	3
5	Chhattisgarh	–	–	–	✓	–	–	1
6	Goa	–	–	–	–	–	–	–
7	Gujarat	–	–	✓	✓	–	–	2
8	Haryana	–	–	✓	✓	–	–	2
9	Himachal Pradesh	–	✓	✓	✓	✓	✓	5
10	Jammu and Kashmir	–	✓	✓	✓	–	✓	4
11	Jharkhand	–	–	✓	✓	–	✓	3
12	Karnataka	–	–	–	✓	–	–	1
13	Kerala	–	✓	✓	✓	–	–	3
14	Madhya Pradesh	–	–	–	✓	–	–	1
15	Maharashtra	–	–	✓	✓	–	✓	3
16	Manipur	–	✓	✓	–	–	–	2
17	Meghalaya	–	✓	✓	–	–	✓	3
18	Mizoram	–	–	✓	–	–	✓	2
19	Nagaland	–	–	✓	–	–	✓	2
20	Orissa	✓	–	✓	✓	–	–	3
21	Punjab	–	–	✓	✓	–	✓	4
22	Rajasthan	–	–	–	✓	–	–	1
23	Sikkim	–	✓	✓	–	–	✓	3
24	Tamil Nadu	✓	–	–	✓	–	–	2
25	Telangana	✓	–	✓	✓	–	–	3
26	Tripura	–	–	✓	–	–	✓	2
27	Uttarakhand	–	✓	✓	✓	✓	✓	5
28	Uttar Pradesh	–	–	✓	✓	✓	✓	4
29	West Bengal	✓	–	✓	✓	–	✓	4
30	Andaman and Nicobar	✓	–	✓	–	–	–	2
31.	Chandigarh	–	–	–	–	–	–	–
32	Dadar and Nagal Haveli	–	–	–	–	–	–	–
33	Daman and Diu	–	–	–	–	–	–	–
34	Delhi	–	–	✓	–	–	✓	2
35	Lakshadweep	–	–	–	–	–	–	–
36	Puducherry	✓	–	–	–	–	–	1

Source Government of India, National Centre for Disaster Management (1999)

4 Disaster Management

As described in the Disaster Management Act 2005, "disaster management means a continuous and integrated process of organizing, coordinating and implementing measures which are necessary and expedient for—(i) prevention of danger or threat of any disaster; (ii) mitigation or reduction of risk of any disaster or its severity or consequences; (iii) capacity building; (iv) preparedness to deal with any disaster; (v) prompt response to any threatening disaster situation or disaster; (vi) assessing the severity or magnitude of effects of any disaster; (vii) evacuation, rescue and relief; and (viii) rehabilitation and reconstruction". Section 35(2) of the DM Act 2005 calls for the central government to ensure that every Ministry of Government undertakes necessary measures of planning and implementing disaster management, as enumerated in Section 36. The idea on contents, objectives and scope of the disaster management plan of a central ministry is given in Section 37.

4.1 Disaster Management and Indian Scenario

India's national disaster management plan, originally approved by the National Executive Committee (NEC) on disaster management in October 2013, was further revised and updated with reference to the adoption of three landmark international agreements in the year 2015 having significant bearing on disaster risk management, i.e.

- Sendai Framework for Disaster Risk Reduction in March 2015;
- Sustainable Development Goals 2015–2030 in September 2015; and
- Paris Agreement on Climate Change at the 2015 Conference of Parties under the United National Framework Convention on Climate Change in December 2015.

While Sendai Framework is the first international agreement adopted within the context of the post-2015 development agenda, the other two agreements have their implications as well in shaping the topology of disaster management plan. One can say that the Sendai Framework has a sharper focus on preventing the creation of fresh risks and places major emphasis on improving the governance for DRR. The Sustainable Development Goals (SDG) also recognize the importance of disaster risk reduction as integral to sustainability, and COP21 Paris Agreement notes the urgent need to take into account the increasing frequency of extreme weather events due to global climate change (National Plan on Disaster Management, India. 2016 [5]).

In addition to it, the post-2015 re-prioritization and the new approaches reflected in these key global agreements, the post-2015 agenda sets the end of 2030 for all the nations to assess the outcomes of their plans and actions in view of realized outcomes.

Therefore, this national plan on disaster management reflects the changes in the outlook and priorities, both nationally and globally. The responsibilities regarding the central ministries/departments are enumerated as follows: (Section 4.4.1 of the National DM Policy) All central ministries/departments will prepare their DM plans including the financial projections to support these plans. The necessary budgetary allocations will be made as part of the Five-Year and Annual Plans. The modalities for the application of these funds will be worked out in accordance with the provisions of the Act. DM Policy's Section 4.4.3 calls for the formulation of plans for mitigation projects at the national level. Central ministries/departments will identify mitigation projects for implementation. The national level mitigation projects are being duly prioritized and approved in consultation with the NDMA.

Disaster management policy of India also recognized climate change as a key aspect contributing to frequency and intensity of hydro-climatic disasters and also increasing the vulnerability of land, ecosystems and socio-economic settings (National DM Policy, Section 5.1.7). Climate change is impacting our glacial reserves, water balance, agriculture, forestry, coastal ecology, bio-diversity and human and animal health. There are clear indications that climate change will enhance the frequency and intensity of disasters like cyclones, floods and droughts in future. In order to meet these challenges in a sustained and effective manner, synergies in our approach and skills for climate change adaptation and disaster risk reduction shall be encouraged and promoted.

The role of central ministries/departments has also been enumerated in disaster preparedness (DM Policy Section 5.2.1). While the national plan will be prepared by the NEC, the disaster and domain-specific plans will be made by the concerned central ministries/departments (Section 5.2.2). The plans prepared by central ministries and departments, states and districts will incorporate the inputs of all stakeholders for integration into the planning process. The participation of all stakeholders, communities and institutions will inculcate a culture of preparedness. A bottom-up approach needs to be adopted for better understanding and operationalization of these plans.

The national policy is also emphasized on the institutional arrangements for disaster risk management (Section 12.2.1 of DM Policy). The entire DM architecture needs to be supported by a solid foundation of frontline R&D efforts, offering sound and state-of-the-art science and technology options in friendly manner. A proactive strategy to enhance mutual reinforcement and synergy among the various groups and institutions working in the area of DM will be recognized. Pooling and sharing of perspectives, information and expertise will be promoted by encouraging such efforts. The identification of trans-disciplinary concerns through a process of "integration" of the talent pool groups will be facilitated and addressed by a standing mechanism at the national and state level. Close interaction with the Central Ministries/Departments of Agriculture, Atomic Energy, Earth Sciences, Environment and Forests, Health, Industry, Science and Technology and Space, and with academic institutions such as the IITs, NITs and universities will be maintained.

5   Need of Data-Centric Planning for Flood Preparedness

With the scenario of country's vulnerability to disaster of any kind in the backdrop, it is highly desirable that the policy formulation to combat the eventualities is backed by evidence-based metrics. In order to foster induction of such practices, it is natural that the country has discrete data sets talking to each other in interoperable manner. Thus, to do so, the available database of all disasters occurring in India can be digitalized and displayed via electronic communication media and the existing information backbone such as SMS, Internet/server, phones, TV and WhatsApp for getting the quick information of different disasters in time for public and practitioners to prepare the people/community in advance to save the society from the impacts of disasters. The management of a disaster is designed according to the new framework of disaster management, and hence, it works in three different phases of a disaster, viz. pre-disaster, during disaster and post-disaster. Here, this paper will discuss about the database planning of flood preparedness, i.e. pre-phase in detail with respect to the present Sendai Framework [6] for disaster risk reduction (DRR).

There are calamitous events on planet earth, and flood is one of them, which disturbs not only the life pattern of the human beings but also destroys their lifelong assets. Floods are caused by high rainfall or natural overflow of dams. The banks of a perennial or seasonal river or a stream flow are markedly higher than the usual as well as inundation of low habitation and agricultural land. Sometimes continuous monsoon rains combine with massive flows from the rivers, and then the floods indeed become disastrous. In India, 22 states and one Union territory (Andaman and Nicobar) are vulnerable to floods. This year (2017), 18 states are badly affected by floods and more than 750 people have already died. As per Ministry of Home Affairs' (nodal ministry) records, the most vulnerable states of India are Uttar Pradesh, Bihar, Assam, West Bengal, Gujarat, Orissa, Andhra Pradesh, Madhya Pradesh, Maharashtra, Punjab and Jammu and Kashmir. On an average, an area of about 8 million hectares (17.50 m ha maximum in 1978) was flooded, of which, on average crop area affected was of the order of 3.302 million hectares (10.15 m ha in 1988). The floods claimed on an average 1464 human lives, and 86,288 heads of cattle die every year, The National Commission on Floods (Rashtriya Barh Ayog) [7]. Government of India (1980) laid great stress on proper flood management for the specific problems of Ganga and the Brahmaputra by adopting a suitable blend of structural and non-structural measures based on long-term strategy inclusive of time and cost effectiveness was evolved to mitigate the flood fury. Floodplain zoning aims to regulate the indiscriminate and unplanned development in the floodplains. It is relevant for both unprotected and protected areas. Hydrological and hydro-meteorological data from 175 flood forecasting stations located in different river basins of the country are collected and analyzed, and then the forecasts are issued by Central Water Commission (CWC) for the benefit of state governments and general public.

Floods present so many problems that no single measure can solve all of them. The basic principle involved in all the methods concerned with mitigation of floods is to *Keep floodwaters away from the man and man away from the floodwaters*. In today's world, even remote corners of river basins have been occupied, forest cover has been systematically destroyed and hilltops have been cultivated to produce more *food* to *feed* the teeming millions. Therefore, the above-mentioned principle has to be amended by and large to suit the present times. It has been seen in recent decades that floods have increased in both volume and frequency either by frequent changes in climate. This has resulted in increased erosion of hills, slopes and riverbanks [8]. As a consequence, the riverbeds have silted up, decreasing the carrying capacity of rivers, and the rivers tend to change their course frequently.

The main aim of non-structural measure is to keep people away from the floodwaters. People who have occupied the flood plains of rivers must be evacuated well in advance to safer zones. If there are any important installations or industrial units located in and around the floodplains, they must be vacated well in time to save them from the damage caused by floods.

Evacuation: The principal task among the emergency measures is the evacuation of the people affected or likely to be affected by floods. These are to be carried out with the support of governments (police, home guards, municipalities, etc.) as well as the non-government organizations (NGOs) related to public health, hospitals, educational institutions, etc. In the flood-prone areas, flood fighting is done by building temporary obstacles and shifting important goods to safer places, so as to minimize the destruction to property. Imparting skill to field workers to rescue marooned people from flood zone is an important constituent in disaster management.

Flood forecasting and warning is now gaining ground and is one of the very scientific methods used to inform the concerned people, well in advance, about impending floods. The past records of rainfall and floods in the river basin are studied by meteorologists and hydrologists and a suitable technique is formulated to forecast floods at various stages of the rivers, 12–48 h in advance, on the basis of antecedent basin conditions, actual rainfall and quantitative precipitation forecast (QPF). At present, there are 175 flood forecasting centres located on the flood-prone rivers of the country, which are manned by experts in meteorology and hydrology. Out of 175 forecasting stations, 147 are for water stage forecast and 28 for inflow forecast used for optimum operation of certain major reservoirs. These 175 stations are located in the flood-prone states with a maximum number in Uttar Pradesh (35) and minimum in Haryana (1) and 4 in Union Territories as displayed in Table 4.

Hydrological and hydro-meteorological data from nearly 100 hydrological and 600 hydro-meteorological stations in these river systems are being collected and analyzed, and then forecasts are issued for the benefit of state governments and general public.

The flood forecasts issued by these centres are disseminated by local administrative officers of the area. Flood warnings are also broadcasted by the local stations

Table 4 Flood forecasting network of CWC (basin-wise FF stations)

S. No.	River-system	Level	Inflow	Total
1	Ganga and Tributaries	77	10	87
2	Brahmaputra and Tributaries	27	–	27
3	Barak system	5	–	5
4	Eastern rivers	8	1	9
5	Mahanadi	3	1	4
6	Godavari	14	4	18
7	Krishna	3	6	9
8	West-flowing rivers	9	6	15
9	Pennar	1	–	1
	Total	147	28	175

Source Central Water Commission, Ministry of Water Resources, Government of India

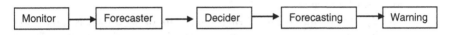

Fig. 2 Components of flood warning system

of All India Radio as well as by TV networks. The final forecasts are then communicated to the concerned administrative and engineering authorities of the states and other agencies connected with flood protection and management telephonically, by fax or by special messengers depending upon local factors like vulnerability of the area and availability of communication facilities. Figure 3 shows the components of flood warning system (Fig. 2).

The modern technique of remote sensing information about floods, supplied by satellites, is available and is being put to use for flood warning. A more important application of the satellite-derived data in the context of flood management is shown below [9].

5.1 Near Real-Time Flood Monitoring and Mapping

The near real limit flood monitoring and mapping involves several facets as detailed below:

(i) **Flood damage assessment**; mapping of existing flood control systems; mapping of river configuration after floods; Identification of eroded or erosion-prone regions; identification of drainage congestion; and

(ii) **Flood-risk zone mapping**. In the coming years, remote sensing technology will play an important role in flood forecasting and flood warning for river basins that are liable to floods. In the course of time, this technique should provide far better information about impending floods than any other method used hitherto. Therefore, floodplain zoning aims to regulate the indiscriminate and unplanned development in flood plains. It is relevant for both unprotected and protected areas (Fig. 3).

(iii) *Flood Proofing*: Such measures help greatly in the mitigation of disasters to the population in flood-prone areas. It is essentially a combination of structural change and emergency action without evacuation. A programme of flood proofing provides the raised platforms for flood shelter for men and cattle and raising the public utility installations above flood levels. Under this programme, several villages were raised in Uttar Pradesh. In West Bengal and Assam, landfills were attempted in villages to keep houses above flood level in some areas. In the Eighth Plan (1992–1997), flood proofing programme was proposed for the Ganga basin states, particularly for north Bihar areas.

(iv) *Flood fighting and Disaster Relief*: On receipt of flood forecasts, the flood forecasting stations (agencies) disseminate flood warnings to the officials concerned and people of the affected area and take necessary measures like strengthening of the flood protection and mitigation works and evacuation of people to safer places etc. before they are engulfed by floods. As a pre-monsoon arrangement, the relief materials must be stocked in advance at appropriate places and distribution measures are initiated to mitigate the miseries.

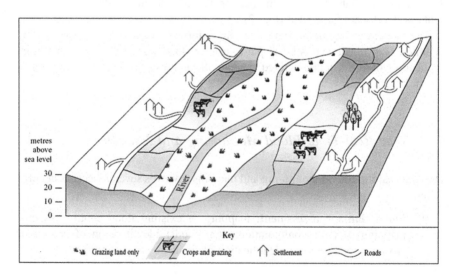

Fig. 3 Floodplain zoning. *Source* Central Water Commission, Ministry of Water Resources, Government of India

(v) *Flood Insurance*: It has several advantages such as it acts as a mean for modifying the loss burden. Albeit, it is being provided to cover the "Flood Risk" yet on a limited and selective scale. This is mainly because of intricacy in the matter of fixing premium and possibility of payment of claims frequently to acutely flood-prone areas. The Ministry of Agriculture has initiated a pilot scheme for crop insurance in the flood-affected areas. In order to mobilize the above, it is of utmost importance to capacitate stakeholders and at the same time create awareness and reach to the unreached. Therefore, training and awareness activities as stated below need to be pursued.

(vi) *Trainings*: At national level, the National Institute of Disaster Management (NIDM) organizes several national training programmes annually on various disasters and on emerging issues related to disaster management and climate change for senior-/middle-level officers of different sectors and states/UTS at its campus and in other states in collaboration with different disaster management cells located at different ATIs for increasing the capacity building of trained persons in the field of disaster management. The significance of such trainings was observed to be useful during the disaster of Chennai flood 2015; when Tamil Nadu Govt. asked NIDM to provide the list of trained officers who had been trained by NIDM in its training programmes on flood disaster management since 2009. NIDM provided the list of around 45 persons trained in the field of flood disaster management to the Tamil Nadu Govt. immediately because NIDM prepares a trainee database, which is uploaded every year after analyzing data of the conducted training programmes on its website [10].

(vii) *Awareness activities*: In schools and colleges, disaster awareness programme should be organized so that in case of flood emergencies, the necessary help and assistance are available without any loss of time. In all flood disaster management and preparedness schemes, the ability to render help to the distressed people in the nick of time has to be kept in view so as to avoid large-scale loss of lives, destruction of properties and communications which, in turn, affect the economy of the country. In addition to it, NIDM conducts the online programmes on disaster management, self-study programmes and satellite training programmes on various disasters for community awareness. IEC materials Do's and Don'ts for different disasters have been uploaded on the websites by NIDM [11] and National Disaster Management Authority (NDMA) [12]. These available data sets cum information are very useful for the preparedness of the society/community against a disaster at national, state, district and local levels. NDMA has prepared some guidelines on flood, cyclones, etc., which are also very useful to disaster managers and community in reducing the impacts of various disaster risks.

(viii) **Response Planning**: The above activities will subsequently lead to response planning. During the south-west monsoon, each year, 25 States/UTs, namely Andhra Pradesh, Arunachal Pradesh, Assam, Bihar, Chhattisgarh, Goa,

Gujarat, Himachal Pradesh, Jharkhand, Karnataka, Kerala, Madhya Pradesh, Maharashtra, Meghalaya, Manipur, Orissa, Punjab, Rajasthan, Sikkim, Tamil Nadu, Tripura, Uttar Pradesh, Uttarakhand, West Bengal and UT of Puducherry reported damage in varying degrees due to heavy rains followed by floods. The economic loss, humans and cattle life loss details are given year-wise from 2007 to 2016 in Table 5 [13].

Every year, India Meteorological Department (IMD) [14] and Central Water Commission (CWC) [15] provide data on south-west monsoon rainfall, flood and other disasters in the form of forecasts and warnings every day to concerned departments and control room of Disaster Management Division of Ministry of Home Affairs, Govt. of India as well as NDMA for taking up the precautionary measures in advance in the forthcoming affected parts of the country. The DM Division (MHA) uploads the state-wise data of flood damage cum forecasting/warning per day on its website as Situation Report (SITREP) during the period of south-west monsoon, i.e. June to September or up to October depending on the pattern of changing climate or weather of the country.

In the present scenario, the planning of response, which is designed before a disaster strikes, includes the updating of relief measures like evacuation, search and rescue operation during emergency, distribution of essential relief materials to the affected people with immediate step and then recovery or restoration of essential services towards normalcy of the life after the disaster. In this way, National Disaster Response Force (NDRF), State Disaster Response Forces (SDRFs), civil defence, Indian Army and other paramilitary forces located in different parts of the country are doing an excellent job by saving the lives of the people. Today, 12 battalions have been set up in different parts of the country. They are well trained, and during zero periods, the persons of NDRF, SDRFs and other paramilitary forces participate in different types of mock drills under the guidance of NDMA. The work done by NDRF was acclaimed globally. Be it the devastating floods in Bihar, UP, MP, Gujarat, Rajasthan and Assam in 2016 and 2017, the Uttarakhand forest fire 2016, the Chennai floods 2015, the Assam Meghalaya floods 2014, the Jammu and Kashmir floods 2014 or disasters like building collapse and train accidents, NDRF has been the most prompt, proactive and visible force to save the people. NDRF has saved more than 5,45,000 lives in various disasters within country and abroad. Apart from its forte of response, the force has actively conducted continuous community capacity-building programmes in the country with a vision to work towards a disaster-resilient nation. NDRF has, so far, trained more than 44 lakh people under various community capacity programs [16]. NIDM also conducts 3–5 training programmes on Indian Response System (IRS) every year for NDRF, SDRFs, Civil Defence, other paramilitary forces and other related sectors for different types of disasters which may occur in the country.

Disaster risk reduction (DRR) is an integral part of sustainable development. For all developmental activities to be sustainable, disaster risk must be reduced. Growing climate change awareness, livelihood and sustainability concerns have brought about a second paradigm shift in disaster management approach

Table 5 Year-wise extent of flood damage during 2007–2016

Year	Population affected	No. of human lives lost	No. of districts affected	No. of villages affected	No. of cattle/ livestock lost	Cropped area affected (in ha)	No. of houses damaged		Estimated value of total damage (Rs. In lakh)
							Fully	Partially/ severely including huts/shade	
2007	592.651	3339	241	52,499	103,341	6,415,288.49	657,262	1,272,390	1,085,346.925
2008	28,090,501	2744	149	19,839	13,114	2,416,669	674,753	917,363	266,837.945
2009	9,574,220	1184	127	12,195	6473	922,779	84,229	258,254	140,004.40
2010	5,659,576	966	76	4975	6290	356,478.1	38,038	71,863	106,586.3
2011	6,932,944	788	75	4423	31,496	1,313,463	112,001	452,866	134,740.1
2012	2,543,175.3	536	310	3611	1,846,223	26,001.63	19,494	70,420	39,965.17
2013	8,529,285.16	1537	363	10,589	110,399	797,969.42	35,875	245,400	472,759.68
2014	8,764,848	1098	91	11,481	3768	60,495.09	9865	124,469	60,660.53
2015	17,307,975	863	50	1733	60,288	1,023,175	133,201 including fully and partially damaged house		132,084.18
2016	11,473,727	970	120	9391	18,815	700,219.79	28,886	114,619	52,876.20
Total average	9,887,684.4111	1402.5	160.2	13,073.6	220,020.7	1,403,253.852	179,360.4	352,764.4	2,491,861.43

worldwide. Now, the focus is shifting from "disaster event" and "minimizing effect of disaster" to "addressing hazards", reducing vulnerability and ensuring sustainability along environment-centric approach [17]. Sendai Framework for Disaster Risk Reduction, 2015–2030 (SFDRR), Sustainable Development Goals, 2030 Agenda (SDGs) and Paris Climate Agreement, 2015 have made the global community realize and recognize that DRR, CCA and sustainable development are linked to each other. Efforts are continuously increasing to adept to climate change and reduce the disaster risks but at the same time the economic and social cost of these disasters are increasing year by year. The theme of Sendai Framework, i.e. "Build Back Better", may be helpful to create a new system of using database electronically for the integration of CCA and DRR to combat the perennial problems of flood, and its resultant destruction has been at the forefront of concern for disaster managers and policy makers.

6 Conclusion

In India, disaster management plans at different levels should be based on scientific data for better preparedness and response. This needs quality data, and its availability to researchers and practitioners at national, state, district, block and village (Local) levels for various hazards such as earthquake, flood, cyclone, drought and others by the concerned organizations functioning in their respective areas, e.g. earthquake by USGS and IMD, rainfall, drought and cyclone by IMD and flood by CWC and IMD. The flood data generated by Central Water Commission (CWC) should be uploaded on a common website or other digital sources through the server of a national agency, i.e. NIC. Similarly, all other available database for other hazards/disasters occurring in India may also be digitalized and displayed via other electronic communication media and existing information backbone such as SMS, Internet/server, phones, TV and WhatsApp for getting quick and right information of different disasters in time so that they may be used as measures of good governance. Indian agencies are using data for simulation, forecasting, providing timely warnings and making disaster management plans. This can certainly minimize economic loss and loss of life and reduce risk to a great extent.

References

1. BMTPC. (2001). A paradigm shift from post-disaster reconstruction and relief to pre-disaster pro-active approach, disaster mitigation and vulnerability atlas of India. Building Material Technology Promotion Council, India.
2. CRED. (2002). EM-DAT: The OFDA/CRED international disaster database. Université catholique de Louvain, Brussels, Belgium. http://www.cred.be/emdat/profiles/natural/banglade.htm.

3. Disaster Management Act (2005). http://www.ndma.gov.in/images/ndma-pdf/DM_act2005.pdf.

4. Timmenmen (1981). http://www.ilankelman.org/miscellany/Timmerman1981.pdf.

5. http://ndma.gov.in/images/policyplan/dmplan/National%20Disaster%20Management%20Plan%20May%202016.pdf.

6. Sendai Framework (2015). http://www.unisdr.org/we/coordinate/sendai-framework.

7. National Commission on Floods (Rashtriya Barh Ayog), 1980, Govt. of India.

8. Uttarakhand Flood (2013). http://nidm.gov.in/PDF/pubs/ukd-p1.pdf.

9. Kulshrestha, S. M. (2000). *Flood management in India*, Joint COLA/CARE Technical Report No. 3, Centre for Ocean-Land-Atmosphere Studies/Centre for the Application of Research on the Environment, Maryland.

10. http://nidm.gov.in/trainee.asp.

11. NIDM. http://nidm.gov.in/default.asp.

12. NDMA. http://www.ndma.gov.in/en/.

13. Flood SITREP(s). http://www.ndmindia.nic.in/.

14. IMD. http://www.imd.gov.in/Welcome%20To%20IMD/Welcome.php.

15. CWC. http://www.cwc.nic.in/.

16. NDRF. http://www.ndrf.gov.in/.

17. http://i-s-e-t.org/resources/training/training-manual-climate-change.html.

Data Marketplace as a Platform for Sharing Scientific Data

Hiranmay Ghosh

1 Introduction

Quality data are one of the major requirements of scientific research in different disciplines. Formulation and validation of scientific theories are activities that essentially rely on many experimental observations. This has been well articulated by Richard P. Feynmann in his lecture: "... *what we're doing here to try to understand nature is to imagine that the gods are playing some great game of chess. And you don't know the rules of the game, but you're allowed to look at the board from time to time, in a little corner, perhaps. And from these observations, you try to figure out what the rules are of the game, what are the rules of the pieces moving.*" Thus, sciences are all about making observations and discovering patterns. For example, after observing several moves in a chess game, we may be able to formulate "*the law for the bishop is that it moves on a diagonal,... (and) that it maintains its color.*" With many more observations, we may discover that there are exceptions to this law, for example when "*a pawn went all the way down to the queen's end to produce a new bishop.*" This explains the need for a very large volume of experimental data that needs to be analyzed to make scientific conclusions. This is true not only for natural sciences like physics and chemistry, but also for quantitative and analytic research in many other disciplines like the social sciences, financial sciences, biological sciences, market sciences, and so on.

While the scientists or research groups have traditionally toiled hard to collect the data for their own scientific experiments, availability of reliable and cheap storage and high-speed data networks in the modern age makes it possible to have a global collaborative approach to generate and utilize data for scientific research. While the basic infrastructure in form of data clouds has been well established over the past few years, there are quite a few technical and motivational challenges that

H. Ghosh (✉)
TCS Innovation Labs, Bangalore, India
e-mail: hiranmay@ieee.org; hiranmay@gmail.com

© Springer Nature Singapore Pte Ltd. 2018
U. M. Munshi and N. Verma (eds.), *Data Science Landscape*,
Studies in Big Data 38, https://doi.org/10.1007/978-981-10-7515-5_7

need to be addressed. We dwell on some of the challenges in the following paragraphs of this article and propose a model for "*scientific data market*" to address them.

2 Challenges in Data Sharing

2.1 Heterogeneity of Data Forms

A major technical challenge is the homogenization of the available heterogeneous data. Since, data are collected independently by different research groups spread across the world; it is not pragmatic to prescribe a uniform way to collect them. The collected data may be in different forms, like numeric, audiovisual and natural language descriptions. The collected data may vary in format, resolution, and accuracy. For example, natural language descriptions can be in different languages; audiovisual data can be in different formats, with different resolutions and different frame rates; and the numeric data may have been collected at different intervals, expressed in different units, and with different accuracies as permitted by the equipment available with the research group. Data transformation and translation services are required to homogenise the data available in different forms and formats.

2.2 Incompleteness, Duplication, and Inconsistency of Data

Different research groups may collect different subsets of data pertaining to a real-world phenomenon, resulting in incompleteness in a dataset. Moreover, same or similar data may be independently collected by different research groups resulting in duplications. A more challenging problem occurs when the data collected by different research groups are inconsistent in the sense that they contain drastically different patterns and lead to different conclusions. Different data cleaning techniques have evolved to address such issues—but more is required to be done in this front.

2.3 Heterogeneity of Context

Data are collected by the different research groups in different contexts. For example, social data collected in different countries may pertain to human beings belonging to different cultural and ethnic backgrounds; market research data may have been collected by different groups may pertain to preferences of people with different genders and age groups; and financial data may have been collected for different economies, and in different countries with different governmental regulations. Thus, context needs to be accounted for while fusing data from multiple sources.

2.4 Heterogeneity of Data Organization

Another major challenge is that the data collected by the different research groups are likely to be differently organized and differently described. In other words, different datasets may be organized around different underlying ontologies. More often than not, the domains may partially overlap and/or a domain may subsume another. Such heterogeneity in data organization makes it difficult to fuse multiple datasets to form a larger dataset. In order to solve the problem, the underlying ontologies of the individual datasets need to be discovered (unless explicitly stated), and the different ontologies need to be aligned, i.e., the relations between the elements in the different ontologies need to be discovered.

2.5 Dynamics of Data

Scientific dataset is not static—new data get generated everyday with new observations being made and new experiments being performed. There need to be a global mechanism to register and discover nascent data. Moreover, new data sometimes make old data obsolete. There should be mechanisms to selectively weed out old data, so that a researcher may not get confused and inadvertently use obsolete data while conducting his or her analysis.

2.6 Data Ownership

While it is good to wish that the data will be freely available within the scientific community for research purposes, we need to remember that data collection often involves tremendous effort and significant expenditure. Besides, access to data provides an edge over other competing research groups. Thus, it is quite natural for a research group to hold on to the data owned by it, unless there is a strong motivation to share.

3   Motivators for Data Collaboration

While technical solutions to some of the challenges mentioned above are in different stages of evolution, we concentrate primarily on the motivational issues in this article. The key challenge is to provide sufficient motivation to research groups to share their data for the research community. A few motivators for a research group to share their data can be as follows.

3.1   Peer Review

More often than not, the research group themselves may draw conclusions based on data gathered by themselves and publish them in form of research papers. Many scientific journals make it compulsory to submit the data together with the research paper for peer evaluation. Such peer reviews add credentials to the scientific research and can be a strong motivator for the research groups to share their data.

3.2   Citation

It is a common practice for a researcher to cite the data used in his or her research publications. This is not only a matter of professional ethics but also required for establishing the credibility for the conclusion drawn. Conclusions made on the basis of analysis of substantial, authentic, and well-known dataset have better acceptability than the conclusions based on some unknown data. The citations can be a motivator for a research group to publish their dataset.

3.3   Data Build-up

Once some data are shared by a research group, others may enhance the dataset and enrich it further. The original contributing group may benefit from such enrichments. Thus, it can be possible to collaboratively build a dataset, much in the same way as many open-source software and public knowledge resources (like the Wikipedia) have been built.

3.4 Government Policies

Since much of the research worldwide is conducted with public funds, sharing of data with peer research labs and educational institutes can be imposed as a precondition for such funding. The projects can be organized in two phases: data collection followed by analysis. An independent assessment of the quality of data collected and *shared* by a research group at the end of the first phase can be imposed a precondition for entering the next (analysis) stage of the project.

3.5 Financial

A research group may have spent significant amount of resources, both money and efforts, in data collection activity. It is quite fair if the cost (at least, a part of it) is recovered by making quality data available to other research groups. The funds so generated can be utilized in further research activities. A counter-argument to such selling may be that it increases the cost of research for the buyer. But, the additional cost will be more than neutralized by the fact that the research group would be required to gather the data by themselves otherwise. Further, this additional cost will be insignificant compared to the overall project cost, where manpower cost is the major driver. Financial gains can also be a strong motivator for commercial organizations to make quality data available to the research community.

4 Data Marketplace

While the practice of selling data has existed for a long time, data marketplace is a relatively new business model and refers to mediation service for data exchange. A few data markets have established themselves in recent years, e.g., Infochimps (infochimps.com), Factual (factual.com), Azure (datamarket.azure.com), and Data Market (datamarket.com), catering to the requirements of the business world. The datasets available on these markets generally bear a price tag; few are available free of cost. The nature of data sold through these services includes business intelligence, advertising, demographics, personal information, research, and market data. The size of the datasets is generally very large, qualifying to be characterized as "big data."

The basic model for a data marketplace is shown in Fig. 1. A data producer deposits data in the marketplace in return of appropriate incentives. A data consumer gets access to aggregated, cleaned, and anonymized data and some data services like retrieval services. Third-party service providers can do value additions on the available data. Above all, they can proactively create a data ecosystem by creating a community of data producers and consumers.

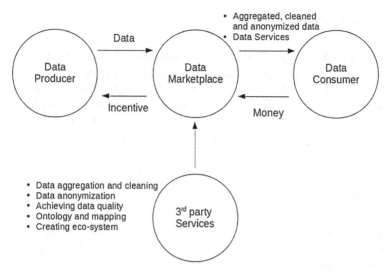

Fig. 1 Model of data marketplace

Establishing a marketplace for scientific data needs certain infrastructure. To begin with, it requires a huge volume of space, where data will be stored. Thanks to the cloud technology, this will be the least of the issues that a data marketplace will encounter. Other technical issues will involve data homogenization, aggregation, and cleaning. It may also offer a platform to host computations closer to the data on the cloud to alleviate need to transmitting large datasets and to prevent unethical proliferation. Discussion fora may be useful for sharing experience in using the datasets. The money received from the data and service consumers can be used to maintain these services. But the biggest administrative challenge will be to create a data ecosystem. Appropriate incentives need to be designed for the data producers. Ensuring data quality and appropriate pricing is necessary to motivate the customers.

The data markets offer many advantages:

1. They offer a long-term nonvolatile storage for large dataset. They also offer appropriate network infrastructure for data access.
2. They provide a platform for discovering and comparing available data. They may also offer infrastructure for user comments and ratings, as well as for a discussion forum. The comments and discussions help a potential user to select quality data appropriate for his use.
3. With datasets growing large, it is often more convenient to move the computations to the data, rather than the other way round. A data market may offer such hosting (cloud) services, which also inhibits unethical proliferation of the data.
4. Availability of datasets on a neutral third-party platform motivates their enhancements and enrichments by others and contributing back.

5. A data market can offer various technical services, e.g., data organization, integration, cleaning, de-duplication, translation, ontology mapping, etc. that enhance the usability of data.
6. It protects the intellectual property rights of the data. It is difficult to *steal* the ownership of a dataset once it is registered in the marketplace. The market administration may also use appropriate legal provisions to safeguard against any unethical use of the data in its custody.
7. They provide a business model for data exchange. They provide an infrastructure for financial transactions and rewarding the data producer. In general, the reward may assume non-monetary forms as well.

5 Conclusion

While data marketplaces are gaining popularity in the business world, it may be worthwhile to experiment with them for scientific data exchange. While financial gains may not be the prime motivator for the scientific communities, the other benefits of the data marketplace can surely be utilized. A mediation service for scientific data exchange, scientific data marketplace has a potential for promoting data sharing and reuse. In Indian context, many a time a diligently collected dataset becomes orphaned once the research project is completed, or the researchers relocate. A data marketplace can provide a permanent home for such data. Data discovery, various technical services, and IPR protections and discussion fora can be the major advantages of a data mediation service for the scientific community.

While the commercial data marketplace is operated by private players, we believe that the government can play a key role in creating a national scientific data marketplace. The marketplace for scientific data can be set up as a not-for-profit organization. While the seed fund for establishing the services may be provided by the government agencies or by other stakeholders, the services can be sustained by generating its own revenue by charging a small fee from its users.

ICSSR Data Service: A National Initiative for Sharing of Social Science Research Data

Jagdish Arora, Pallab Pradhan, Yatrik Patel, Miteshkumar Pandya, Hiteshkumar Solanki and Divyakant Vaghela

1 Introduction

The concept behind open data is free availability of data to everyone for use, reuse and republish without restrictions of copyright or other control mechanisms. The objectives of the open data movement are perfectly aligned to other "open movements" such as open source, open access, open standards, open hardware and open content. The philosophy behind open data has long been established, but the term "open data" itself is recent, gaining popularity with the rise of the Internet and World Wide Web and, especially, with the launch of open data government initiatives such as www.data.gov (USA), data.gov.in (India) and data.gov.uk (UK).

Data has its own value as an output of research. It contributes to the enhancement of profiles of contributing researchers, departments and institutions. Data is the key to scientific analysis of research as well as for furthering research. It substantiates the basis of research, validity of inferences drawn and decision making. The better the quality of data, in terms of both precision and statistical

J. Arora (✉) · P. Pradhan · Y. Patel · M. Pandya · H. Solanki · D. Vaghela
Information and Library Network (INFLIBNET) Centre, Gandhinagar 382007, Gujarat, India
e-mail: jarora42@gmail.com; jarora@inflibnet.ac.in

P. Pradhan
e-mail: pallab@inflibnet.ac.in; ppradhan86@gmail.com

Y. Patel
e-mail: yatrik@inflibnet.ac.in; yatrik@gmail.com

M. Pandya
e-mail: mitesh@inflibnet.ac.in

H. Solanki
e-mail: hitesh@inflibnet.ac.in

D. Vaghela
e-mail: divyakant@inflibnet.ac.in

© Springer Nature Singapore Pte Ltd. 2018
U. M. Munshi and N. Verma (eds.), *Data Science Landscape*,
Studies in Big Data 38, https://doi.org/10.1007/978-981-10-7515-5_8

significance, the better the analytical output and inferences made. If the data is made available more openly and freely, it enables better reuse and analysis on the one hand and evaluation of the quality of the data on the other [3].

There has been a persistent demand across the globe for open availability and sharing government and research data collected or developed through public funding. Data repositories offer an opportunity to relook, reanalyse and rework on the data collected painstakingly by researchers and government organizations. Funding agencies are increasingly mandating submission of data into data repositories for the larger benefit of academic community.

The article elaborates on background, genesis, implementation strategy and current status of the ICSSR Data Service: Indian Social Science Data Repository. It elaborates on organization of data in repository, features and functionality and data sets available in ICSSR Data Service. Article enumerates on "Explore Online" and "ICSSR Data Analytic" tools in detail.

2 Data Sharing: Justifications and Methods of Data Sharing

2.1 Justifications

Data sharing is the practice of making data used for scholarly research available to other investigators/researchers for reuse, reinterpretation and verification. Several funding agencies, institutions and journals have well-defined policies on data sharing with an aim to promote transparency and openness. Data sharing is increasingly considered as an important part of scientific process. The real value of research data lies in its use, reuse and revalidation. In addition, publicly funded research data are produced in the public interest and, therefore, should be shared widely. Several funding agencies insist that the data generated from the research project funded by them should be deposited in a specified data repository or in a repository that is publicly accessible. Likewise, several journals have designated their preferred data repository for submission of data related to articles published in their journals.

Some of the important justifications for data sharing through data repositories are as follows:

(i)   Availability of data through data repositories enhances its impact and visibility as well as its reuse for further scientific and educational purposes;
(ii) It enhances the profile and recognition of the researchers, institutions and funding agencies;
(iii) Sharing of data maximizes the possibility of data transparency and accountability for a completed research project as well as its research output;
(iv) It allows other researchers to understand, evaluate and advance a piece of research while relying on an individual/group of individuals or organization's specific work, and thereby providing them an opportunity to collaborate in further research on the same topic;

(v) Data repository enables collection and organization of data from diverse sources and supports its search, browsing and discovery. In addition, data repository provides a powerful research environment for analyses and visualization of related datasets from diverse sources through intuitive interfaces that may either be inbuilt into the repository or use a variety of third-party analytical tools;

(vi) Data repository provides simpler, unified and seamless access to an extensive and expanding range of high-quality scientific and social research data on varied subjects in an open environment with an aim to meet the needs of academic researchers and business analysts;

(vii) Primary data available in the repository can be downloaded, analysed and reported in a consistent format using off-the-shelf web-based or desktop-based software. Data repositories facilitate simplified report development using standard table, column and field naming conventions;

(viii) It facilitates manipulation and graphical depiction of selected data using enhanced visualization tools; and

(ix) The software and backend server for data repository provide multiple layers of security features. As such, access can be restricted to specific tables and/ or columns or rows in a table.

2.2 Methods of Data Sharing

Researchers can share their data using one of the following three methods.

2.2.1 Data Repositories

Data repositories generically refer to a central place where data is stored and maintained often for safety or preservation. It can be a place where multiple databases, datasets or files are located for preservation and distribution over a network that is directly accessible to the user. It may include requisite infrastructure often referred as data archives or data centres required for obtaining and depositing data to facilitate further sharing, analysis and reuse. Data repositories are continuing to evolve in almost all disciplines as an active component of virtual research ecosystem.

2.2.2 Institutional Repositories

Usually based in an academic institution, institutional repositories store and share data created by that institution's faculty members, staffs or students. The software used for institutional repositories is generally not suitable for hosting datasets.

2.2.3 Self-preservation

Self-preservation is an option for individual researcher to share their data through their own websites or by publishing it as a journal article or discussion forum.

However, Indian academic and research institutions have a lot of catching up to do as compared to their western counterparts as far as open data and data sharing are concerned. The International Federation of Data Organizations (IFDO) in their report entitled "Policies for Sharing Research Data in Social Sciences and Humanities" published in 2013 on data policies of 32 countries found that the Northern American and Western European countries were best represented in terms of responses received in the online survey conducted by IFDO. Questions were focused on whether key research funders had formal data sharing policies and if they did, what kind? Response alternatives provided were as follows: (i) the funder requires data sharing; (ii) the funder recommends data sharing; or (iii) the funder has no requirements or recommendations related to data sharing.

The survey concludes that only a handful of funding agencies in a very few countries have adopted the policy of data sharing as a mandate. Only a handful of funding agencies in a few countries have implemented data sharing as mandatory requirement for research conducted in the projects awarded to individuals and institutions funded by them. In UK, the Economic and Social Research Council (ESRC) has mandated that all projects funded by ESRC should archive the data generated in projects funded by them at the UK Data Service. Such mandates are becoming common in the USA [1].

In the year 2012, the Ministry of Science and Technology, Government of India, initiated a draft on data sharing policies entitled "National Data Sharing and Accessibility Policy (NDSAP)" to share non-sensitive data available either in digital or in analogue forms that are generated using public funds by various ministries/departments/subordinate offices/organizations/agencies of government of India as well as States. The NDSAP policy is designed to promote data sharing and enable access to government of India owned data for national planning, development and awareness [2].

3 ICSSR Data Service: Background, Genesis, Objectives and Stakeholders

3.1 Background

Established in 1969, the Indian Council of Social Science Research (ICSSR) is an apex body of government of India who sponsors, promotes, monitors social science research and provides financial help to carry out research in the country. Presently, ICSSR is assisting 27 research institutes and 6 regional centres across India. A large volume of data is being generated by these research institutes as well as by individual researchers through various research projects funded by the ICSSR. Besides, Ministry of Statistics and Programme Implementation (MoSPI) collects a comprehensive set of statistical data sets generated by large-scale government surveys, i.e. National Sample Survey (NSS) conducted by the National Sample Survey Office (NSSO) and Annual Survey of Industries (ASI) conducted by Industrial Wing of the Central Statistics Office (CSO). Likewise, all ministries and departments of government of India generate and collect statistical data on their activities and services. Initiative to set up social science data repository was taken by the ICSSR with an aim to make all social science statistical datasets generated by all government and non-government initiatives, available in open access to entire social science research community not only in India but also all over the world and to foster substantial social science research in the country.

3.2 Genesis

A committee of experts on data repositories and a sub-committee for setting up "ICSSR Data Centre" were set up by the ICSSR to initiate the process of setting up of ICSSR Data Service. The committee, in its report entitled "Road Map for Setting-up a Data Repository @ ICSSR and Sequence of Activities and Timeline for Quick Implementation", submitted a detailed implementation plan for setting up of data centre with requisite computing infrastructure, options for data repository software, data sourcing, data acquisition policy, preprocessing, hiring of manpower, launching of services, etc. Selected members of the committee were deputed to attend a workshop on *"Underpinning the Development Data Revolution: Collaboration in Data Archiving"* held at University of Essex, UK, on 7 and 8 January 2015. The workshop was funded by the UK Research Council as per the MoU signed between ICSSR and UK Research Council. The workshop not only imparted training on various aspects of setting up and maintenance of data repositories, but also facilitated formal and informal interactions between participants and experienced faculty having decades of experience of maintaining data repositories. Every participant who attended the above-mentioned workshop gave their technical feedback for setting up the ICSSR's Data Repository based on knowledge gained at the workshop. Those feedbacks were compiled, reviewed and taken in consideration for their implementation towards the development of repository.

3.3 Objectives

The first and foremost objective of ICSSR Data Service is to provide access to a wide range of datasets generated by the MoSPI, New Delhi, social science institutions under direct purview of ICSSR and other government organizations, and to researchers who are looking for high-quality social and economic research datasets. The basic aims and objectives of the repository are as follows:

- To serve as a national data service for promoting powerful research environment through sharing and reuse of data amongst social science community in India;
- To acquire, process, organize, preserve and host research data along with its metadata with ETL (extract, transform and load) facilities of raw data in social sciences and related domains collected from diverse sources for easy sharing and access;
- To facilitate online submission, access, search, browse, discovery, conversion, analysis and visualization of data through intuitive interfaces;
- To impart training and spread awareness about benefits of data sharing and reuse amongst social science research community in India; and
- Interact, cooperate and collaborate with other national and international data services and repositories for data and resource sharing and improved management of data services.

3.4 Stakeholders

The ICSSR Data Service is a national-level data repository service for the social sciences to facilitate data sharing, preservation, accessibility and reuse of social science research data collected from entire social science community in India. The repository of the ICSSR Data Service is developed based on participatory approach. As such, while the Ministry of Statistics and Programme Implementation (MoSPI) contribute unit-level data in the repository, it is expected that other ministries, government departments, organizations, students, researchers and faculty would also contribute their research data into the repository and other ministries, government departments and organizations subsequently. Besides, individuals and organizations are also welcomed to contribute secondary data driven from unit-level NSS and ASI data sets and make them available for the benefit of other researchers. Major stakeholders of the ICSSR Data Service are as follows:

- Ministry of Statistics and Programme Implementation (MoSPI) and its two organs, namely National Sample Survey Organization and Central Statistics Office (CSO);
- ICSSR: Indian Council of Social Science Research (ICSSR) and its 27 constituent research centres located across the country;
- Other ministries, government departments and policymakers as users as well as contributors;
- Students, researchers and faculty as users as well contributors;
- Working professionals and NGOs as users as well as contributors;
- Universities and colleges as organizations that use and contribute to ICSSR Data Service and define policies on data generation and its delivery;

- Foreign users: students, research scholars, scientists and faculty members from institutions abroad with which ICSSR has bilateral understandings/agreements on sharing of resources subject to the condition that similar facilities will be reciprocated by such institutions with respect to resources held by them; and
- Any others individual/independent researchers, government organizations, private firms, NGOs and any other institutions working on social sciences and related domains who wish to deposit their data into the repository by complying to the data repository policies.

4   ICSSR Data Service: Current Status and Organization of Data Repository

4.1   Current Status

The "ICSSR Data Service" is the culmination of signing of Memorandum of Understanding (MoU) between Indian Council of Social Science Research (ICSSR) and Ministry of Statistics and Programme Implementation (MoSPI). The MoU provides for setting up of "Social Science Data Repository" and hosting of NSS and ASI data sets into the data repository. The Information and Library Network (INFLIBNET) Centre, Gandhinagar, was assigned the task of setting up of the data repository along with all related activities as an ICSSR-sponsored project considering the facts that ICSSR does not have infrastructure and expertise required for setting up, maintaining and hosting the data repository.

After exploring a number of open source softwares available for data repository including CKAN, DKAN, NADA and Dataverse, finally the implementation team at the INFLIBNET Centre zeroed down to NADA as the most suitable platform to host the ICSSR's Data Repository. NADA software was customized as per the requirement of ICSSR Social Science Data Repository. A number of existing social science data repositories around the world were explored with an aim to incorporate important features and functionalities into ICSSR Data Service. The NADA software was installed on a server, especially acquired for this purpose in the Data Centre at the INFLIBNET Centre. The ICSSR Data Service hosts a comprehensive set of statistical datasets generated by large-scale government surveys, i.e. National Sample Survey (NSS) conducted by the National Sample Survey Office (NSSO) and Annual Survey of Industries (ASI) conducted by Industrial Wing of the Central Statistics Office (CSO). Subsequently, scope of the repository would be expanded to include datasets from all social science institutes under ICSSR's direct purview, other organizations, NGOs and individuals as well as government agencies. The ICSSR Data Service is designed to provide seamless, integrated access and support to meet the requirements of academic researchers as well as policymakers who heavily rely on high-quality social and economic data for their research.

Welcome to ICSSR Data Service

Fig. 1 "Home" Page of ICSSR Data Service

The ICSSR Data Service was launched formally on 20 June 2016 by Dr. T.C.A. Anant, Secretary and Chief Statistician of India, MoSPI, at ICSSR, New Delhi. Figure 1 is the screenshot of the "Home" page of ICSSR Data Service.

4.2   Data Sets in the ICSSR Data Service

Currently, 136 data sets provided by the MoSPI are extracted, transformed and hosted on the repository with their necessary metadata elements. The datasets in the repository are organized into eight major thematic categories defined by the MoSPI. Moreover, HASSET (Humanities and Social Science Electronic Thesaurus) developed by UK Data Service is being used for assigning key terms to the data sets. The INFLIBNET Centre has signed a licensing agreement with the UK Data Service, University of Essex, UK, for using HASSET thesaurus. The thesaurus is being integrated with metadata input worksheet of the data repository to facilitate assigning of key terms at the time of keying-in metadata for each data set. Moreover, authority files of names, places and codes are being created based on the data sets received from the MoSPI. The Centre is also adding country-specific key terms to the thesaurus for indexing data sets and other social science literature emanated from India.

For the benefit of data users, the Centre has developed online data analytical and visualization tools that support mapping and geo-coding applications using "R" language, which is inbuilt in the data repository itself.

4.3   Organization of Data in Repository

All the data sets available in the ICSSR Data Service are organized according to their data characteristics. The datasets are categorized into eight major categories, namely Debt and Investment, Domestic Tourism, Education, Enterprise Survey, Employment and Unemployment, Housing Condition, Household Consumer Expenditure and Healthcare. Users can search these data sets using intuitive search interface of the repository to find and locate their desired data sets or information related to the data sets. Users can also search the Data Catalogue that facilitates users to search by title, creator, keywords, description, subject, coverage, series or identifier. After finding their desired data, users may analyse the data online, export/ download datasets for further self-analysis or else may download the excerpts of their online analysis. Visualization tools, mapping and geo-coding applications are inbuilt in the data repository through Data Analytic Tool.

Fig. 2 Display of datasets in "Microdata Catalog"

Fig. 3 Organization of a dataset in ICSSR Data Service

All data sets available in the repository are available under "Microdata Catalog" in Central Data Catalogue available at the "Home Page" of the ICSSR Data Service which is shown in the screenshot in Fig. 2.

Users are required to "Register" before they are allowed to download data sets from the repository. Once registered, a user can log into his/her account for downloading data sets from the repository. Users are redirected to the "Central Data Catalogue" of the ICSSR Data Service where complete list of data sets available in the data repository is displayed. Each dataset available in the repository is organized under the following four tabs: (i) Documentation; (ii) Study Description; (iii) Data Description; and (iv) Get Microdata, which are depicted in Fig. 3.

4.3.1 Documentation

"Documentation" contains detailed questionnaires, technical documents, other accompanied materials used in the survey process and reports that describe the key results of the study. User can view/download these materials and reports for their use.

4.3.2 Study Description

"Study Description" provides details about the study which include its metadata, identification, abstract, sampling procedure, data collection methodology, time period, ease of data access and export data in DDI-XML format.

4.3.3 Data Description

"Data Description" provides an elaborate description on data which includes the block-level/file-level description, number of cases and variables available in each block/file. A user can see details on each block to find out variables, number of variables, variables' name and their associated label.

4.3.4 Get Microdata

Users can directly download the raw data files submitted by the Data Depositor or the transformed data files "TSV Transformed Data" or "SAV Files of Data" of the raw data by the ICSSR Data Service. The user can also see the volume and size of those raw data and transformed data. Users have an option to download the dataset in a number of formats such as xls, Jason and Stata and explore it at their own convenience using third-party analytical tools.

4.4 ICSSR Data Service: Important Features and Functionalities

Built on NADA software platform, the ICSSR Data Repository supports following features and functionalities:

- Supports search and discovery through elaborate metadata description of each data set using Data Documentation Initiative (DDI) Metadata Standard (DDI-MS);
- Provides raw as well as transformed data in multiple formats;
- Provides options for online analysis through inbuilt data analysis tool by choosing multiple variables;
- Generates cross tabulation for various datasets;
- Provides for data visualization through bar charts, line and scatter diagrams, pie charts, stacked charts, histogram, etc.;
- Supports multiple output formats such as CSV, TSV, PDF, EXCEL, DTA and SAV;
- Multiple search and browse options; and
- Visualization of geo-coded data on maps available online.

5 Data Selection, Acquisition, Processing and Withdrawal: Policies and Practices

5.1 Data Selection

The scope of ICSSR Data Service is to accommodate data collected from social science and other related subject domains in a computer processable format. All data and data-related descriptions will be in English till capabilities are developed for processing data in other languages in the data repository. The ICSSR Data Service determines the status of data to qualify for inclusion in the data repository on the basis of their:

- Demonstrated importance, relevance, completeness and availability of metadata;
- Quality of data based on agency reputation, research and public policy outcomes, comparability and timeliness; and
- Easy access conditions to facilitate faster acquisition procedures.

The ICSSR Data Service considers and accepts mainly: (i) raw or preliminary data; (ii) data that are ready to use and ready for full release; (iii) unit-level summary data; and (iv) tabulated, analysed and derived data for inclusion in the repository.

5.2 Data Acquisition

While the ICSSR has signed an MoU with the Ministry of Statistics and Programme Implementation for getting NSS and ASI data sets on regular basis, it would be imperative for the ICSSR to source data from all other sources for the benefit of entire social science community in India and abroad. As such, ICSSR would identify and tap various sources of data and ways and means for its systematic acquisition from individuals, private and government organizations, educational institutions and NGOs. ICSSR may sign MoU with other ministries/government departments/NGOs/universities/other academic institutions.

Besides other means of data acquisitions, (i) ICSSR Data Service would have an interface that would facilitate individuals and organizations to upload data available with them in the repository; (ii) it would be obligatory for individual researchers who approach ICSSR's Data Processing and Analytics Unit to deposit their raw data as well its derivatives into the data repository @ ICSSR; and (iii) ICSSR may consider mandating submission of data sets created by its own institutions in the repository as well as under the ICSSR-funded projects. An expert committee consisting of eminent social scientists is being set up at ICSSR for identifying sources of data and method of its acquisition.

5.3 Data Processing and Metadata Schema

Standard metadata is necessary for every dataset to facilitate its search, discovery and browsing. In general, metadata provides context for data and helps researchers and students to distinguish one dataset from another or related datasets available in a data repository. Metadata is often structured as a series of fields and recorded in XML format. Most repositories get the metadata created by its submitter through online form that makes it easy to create and upload metadata. Metadata for data sets are often structured using international standards or schemes such as Dublin Core, ISO 19115 for geographic information, Data Documentation Initiative (DDI), Metadata Encoding and Transmission Standard (METS) and General International Standard Archival Description (ISAD(G)) [5].

The ICSSR Data Service employs DDI-XML-based descriptive metadata schema for assigning descriptive metadata to data sets deposited into the repository which is a natural choice as a common metadata standard for the social science data infrastructures around the world. Also, keywords are assigned to the data/data sets available in the ICSSR Data Service using Humanities and Social Science Electronic Thesaurus (HASSET) developed by UK Data Service.

5.4 Data Types and Formats

The ICSSR Data Service supports varieties of preferred data formats for hosting into the repository which are machine readable. The ICSSR's Data Repository mainly holds unit-level statistical data of National Sample Survey (NSS) conducted by National Sample Survey Office (NSSO) and Annual Survey of Industries (ASI) conducted by Industrial Statistics (IS) Wing of Central Statistics Office (CSO) along with their supporting documents, i.e. questionnaires, data collection methods, codebooks and project summaries/descriptions. The unit-level NSS and ASI statistical data are available mostly in Excel, CSV, SPSS and STATA formats, while the supporting documents, i.e. questionnaires, data collection methods, codebooks and project summaries/descriptions are in plain text, Word, PDF and HTML formats. However, users have the liberty to export/download, convert and use the data sets using different statistical software packages according to their requirements.

5.5 Data Version Control

It is necessary to keep various versions of a dataset along with the original deposited dataset in the repository as dataset can be updated, edited, revised and deleted. ICSSR Data Service adheres to this to ensure the originality and authenticity associated with a data set by following the given steps:

- The repository uses numbers for all versions of a dataset as per its updation, edition, revision and also with its status, i.e. draft, interim and final, to clearly indicate their users.
- The same details may be recorded in DDI metadata record with the data file.
- The repository always keeps the original deposited datasets along with its supplementary information.

- A dataset is being kept in different formats along with its original within the repository.
- The repository uses persistent identifiers to cross-link and cross-reference different versions of a dataset with the original.

5.6 Data Assessment and Modification

Acceptance of data may not compulsorily be guaranteed after submission as the data have to be internally evaluated by the data experts from ICSSR Data Service. Data/datasets proposed for deposit are reviewed for eligibility of the depositor, relevance to the scope of the collections, valid data formats, etc. Further, the submitted data would go for an internal data quality check to ensure that the quality standards are maintained, i.e. the accuracy, its consistency, documentation, metadata, free from any sort of legal issues and privacy of individuals, and do not compromise with the national security. The datasets along with the data files are duly checked and validated to ensure that variables and values are accurate according to the documentation supplied by the depositor and are sufficiently labelled for secondary use. The submitted data would qualify for deposition into the repository only after thorough examination.

The ICSSR Data Service extracts and transforms the data from raw datasets before uploading it to the data repository with necessary metadata and documentation for the benefit of researchers. Besides hosting unit-level datasets, the ICSSR Data Service also provides access to secondary datasets derived from the unit-level datasets on different parameters. Training materials and guidance to meet the needs of data users, owners and creators are also offered through this platform.

5.7 Data Withdrawal

The datasets deposited into the repository may be withdrawn or removed from the repository in later phase by the ICSSR Data Service for issues involving violation of copyright, legal requirements and proven legal violations, data involving national security, falsified data and unfair research and confidentiality concerns. However, the data owner or depositor may also request to the ICSSR Data Service for withdrawal/removal of his/her deposited data from the repository in unavoidable and desired conditions.

6 Data Access and Reuse

The ICSSR Data Service promotes the philosophy of open access and adheres to open data policy. The ICSSR endorses the philosophy that "Digital research data should be easy to find, and access should be provided in an environment which maximizes ease of use, provides credit for and protects the rights of those who have gathered or created data, and protects the rights of those who have legitimate interests in how data are made accessible and used [4]".

Datasets and their accompanying materials such as survey questionnaire, support guides and instructions to data collectors are publicly available in the ICSSR Data Service. Users are free to search, browse, access and reuse these datasets for further study, research, teaching and learning. Users are required to register with ICSSR Data Service before downloading the datasets or analyse it online wherever possible. Users should not breach the intellectual property rights of the data owner while reusing the data from the ICSSR Data Service. A user manual on "How to Search and Access Datasets" is available at: http://www.icssrdataservice.in/files/ICSSR_Data_Access_Manual.pdf.

There are no restrictions or limitations on access and use/reuse of data for fair and personal use. However, restrictions may be applicable on access and reuse of some specific datasets due to certain circumstances. The ICSSR Data Service facilitates access to each datasets deposited in the repository in accordance with the request and conditions specified by the data owners at the time of depositing dataset and signing license agreement, or any other considerations deemed important by the ICSSR Data Service itself. The following three types of access control can be exercised on the datasets available in the repository:

- **Open Data**: Access to open data generally means that data generated through public-funding should be freely available in open access in easy, timely, user-friendly and web-based methods without any process of registration/authorization to one and all without any restrictions. User can freely download these data directly from the repository.
- **Safeguarded Data**: Data can be accessed only to the registered users only after a prescribed online registration process or authorization by the ICSSR Data Service once they agree to the terms and conditions of data usage policy displayed to them at the time of request.
- **Controlled Data**: Access to these kinds of data would be controlled through Secured Lab, stored in a secured server. Different versions of same datasets may be derived for its open and controlled access. Data declared as sensitive and highly confidential, by government of India policies, will be accessible only through this mode.

Access to and reuse of metadata records of the datasets available in the data repository are free of charge and openly accessible to all users for teaching, research and analysis, but not for any commercial purposes. However, access to some of the metadata may be controlled or restricted by the ICSSR Data Service. Proper attribution and citation are required from the users after use/reuse of metadata.

7   Data Analysis and Online Exploration: "Explore Online" and "ICSSR Data Analytic Tool"

There are two inbuilt options available in ICSSR Data Service, namely "Explore Online" and "ICSSR Data Analytic Tool" which facilitate online exploration, analysis and visualization of datasets (Fig. 4).

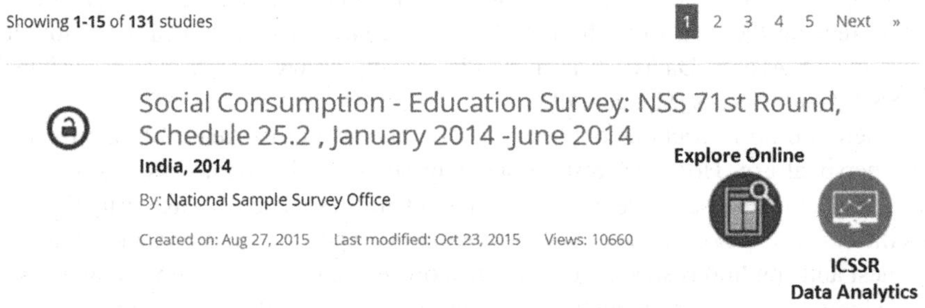

Showing **1-15** of **131** studies   1   2   3   4   5   Next   »

Social Consumption - Education Survey: NSS 71st Round, Schedule 25.2 , January 2014 -June 2014

India, 2014   **Explore Online**

By: National Sample Survey Office

Created on: Aug 27, 2015   Last modified: Oct 23, 2015   Views: 10660

ICSSR
Data Analytics

Fig. 4 Display of "Explore Online" and "ICSSR Data Analytic Tool" options

7.1   Explore Online

User can explore the datasets online by opting for "Explore Online". In this option, user can explore the data online and visualize results as charts, tables, graphs and map using techniques and tools inbuilt into the data repository. Further, user also has an option to select preselected data from the drop-down menu/list for their graphical representation along with its tabular format.

7.2   ICSSR Data Analytic Tool

User can explore a dataset online through "Explore Online" option as mentioned above or can analyse and visualize the data using "ICSSR Data Analytic Tool". The "ICSSR Data Analytic Tool" is developed in "R" language for advance analysis and visualization of datasets available in the ICSSR Data Service. User can perform a number of analyses by using ICSSR Data Analytics such as univariate analysis, data transformation, cross tabulation, pivot analysis and generation of various maps and charts (Fig. 5).

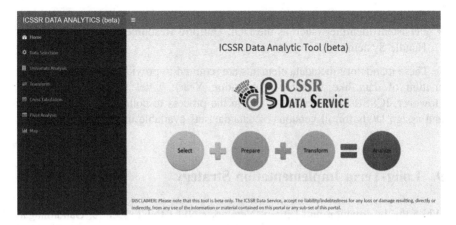

Fig. 5 Dashboard of ICSSR Data Analytic Tool (beta)

8 Data Citation and Principles

Citation is a significant feature of any research publication. It is a healthy academic and research practice that provides information regarding sources of data used in a given work. Bibliographic information provided in citations helps other researchers to identify and locate the referenced data. Further, tracking of citations to specific data or collections helps to determine the level of usage and its impact. However, present citation practices do not put much emphasis on data citation and its attribution as compared to other evidence and sources like research articles, books and patents. Proper practice of data citation would help to further advance the practice of data sharing and reuse.

The ICSSR Data Service expects that users of data from the repository should provide correct citation and acknowledgement for data used by them.

For Example:

Central Statistics Office (Industrial Statistics Wing), MOSPI, Government of India. Annual Survey of Industries: 1983–84 [computer file]. × Edition. Kolkata, India: ICSSR Data Service [distributor], October 2012. SN: ×, DOI/Handle: xxx xxx xxx.

ICSSR Data Service follows "DataCite" metadata structures for citation of data or datasets:

- Identifier: A unique string that identifies a resource—the allowed value is the DOI.
- Author/Creator: The main researchers involved in producing the dataset.
- Title: A name by which the resource is known.
- Publisher: Name of the publisher of the dataset.
- Publication Date/Year: Year in which the data was or will be made publicly available.

- Version: Version of data.
- Persistent Identifier (such as the DOI, Uniform Resource Name URN/URL or Handle System).

These mandatory metadata elements are required to provide the elements for the citation of data like "Creator (Publication Year): Title. Publisher. Identifier". However, ICSSR Data Service is still in the process to collaborate with DataCite and assign DOIs for all versions of data/datasets available in the data repository.

9 Long-Term Implementation Strategy

While the Information and Library Network (INFLIBNET) Centre, Gandhinagar, was assigned the task for setting up of the data repository along with all related activities as an ICSSR-sponsored project, steps are being taken to involve staff, officers and fellows of ICSSR with an aim to transfer the responsibility of maintenance of repository gradually to the ICSSR. As such, the following facilities are being set up at the ICSSR:

9.1 Centre for Data Acquisition and Management @ ICSSR

While ICSSR may continue to host its data repository at INFLIBNET Centre, it is imperative for the ICSSR to set up a Centre for Data Acquisition and Management @ ICSSR which will be responsible for data acquisition and its preprocessing prior to hosting. The ICSSR has identified staff members from NASSDOC who would be responsible for taking up the activities involved in data acquisition and management. ICSSR may also appoint a senior consultant and associated staff required for handling the activities of this Centre.

9.2 Setting up of Data Repository Development Centre @ ICSSR

While ICSSR may continue to use the hosting facilities provided by the INFLIBNET Centre for hosting its data repository, it would be imperative that internal capacities are developed in-house in long run for maintenance and development of the repository. As such, Data Repository Development Centre is being set up at the ICSSR that would consist of a development server and five PCs. While the server would host a replica of data repository at INFLIBNET Centre, the PCs would enable ICSSR staff and fellows to interact and manage the Data Repository at INFLIBNET Centre. This infrastructure could also be used for processing and analysis of data before it is hosted in the Data Repository. The hardware for data repository at INFLIBNET Centre was acquired and installed, and orders for server and PCs for Data Repository Development Centre at the ICSSR have been placed.

Training was imparted to the staff members from ICSSR at the INFLIBNET Centre in handing of ICSSR Data Service.

10 Conclusion

The practice of making research data open and its sharing is not commonly practiced in Indian academic and research environment. Availability of NSI and ASI datasets through ICSSR Data Service may become a trend setter which will have long-lasting impact on scholarly community in India. ICSSR Data Service is a major step taken by the ICSSR in this direction to realize the potential of open data sharing in social science domains. As a policy, the ICSSR Data Service promotes data sharing to encourage reuse of data and provide information on developing and generating social science research data and its management.

References

1. Kvalheim, V., & Kvamme, T. (2014). IFDO: Policies for sharing research data in social sciences and humanities: A survey about research funders' data policies. www.ifdo.org.
2. National Informatics Centre (NIC), Government of India. (2014). Implementation guidelines for national data sharing and accessibility policy (NDSAP) (Ver. 2.2). https://data.gov.in/sites/default/files/NDSAP_Implementation_Guidelines_2.2.pdf.
3. Ramachandran, R. (2000). Public access to Indian geographical data. *Current Science, 79*(4), 450–467.
4. Research Information Network (RIN). (2008). Stewardship of digital research data: a framework of principles and guidelines. http://www.rin.ac.uk/system/files/attachments/Stewardship-data-guidelines.pdf.
5. UK Data Archive (UKDA). (2011). Managing and sharing data. http://www.data-archive.ac.uk/media/2894/managingsharing.pdf.

Reading List

Green, A., Mcdonald, S., & Rice, R. (2009). Policy-making for research data in repositories: A guide (Ver. 1.2). www.disc-uk.org/docs/guide.pdf.
Inter-university Consortium for Political and Social Research (ICPSR). (2012). Guide to social science data preparation and archiving: Best practice throughout the data life cycle (5th ed.). Ann Arbor, MI. http://www.icpsr.umich.edu/files/deposit/dataprep.pdf.
Consortium of European Social Science Data Archives (CESSDA). http://cessda.net/CESSDA-Training.
DataCite. https://www.datacite.org/.
ICSSR Data Service. http://www.icssrdataservice.in.
UK Data Service. http://ukdataservice.ac.uk/.

Prismatic Consumer Insights Through Big Data: A Case Study of National Consumer Helpline

Suresh Misra and Deepika Sur

1 Introduction

In the digital age that we are living in, data is a term often used. Data is defined as pieces of information, of a specific kind, and in specific formats. Any computer software is made up of two units—the algorithm (set of instructions designed to form a programme) and the data. Data can exist in a variety of forms—as numeric or alphanumeric. A collection of data that is organized and collated in the form of tables, fields, etc., are called a database. Databases are used for analysis and findings and are used where typically information is required. Today big data is being increasingly used to better understand customers and their behaviours and preferences. Companies are keen to expand their traditional data sets with social media data, browser logs as well as text analytics and sensor data to get a more complete picture of their customers.[1] The big objective, in many cases, is to create predictive models.

The story of how data became big started many years before the current buzz around big data.[2] Big data refers to technologies and initiatives that involve data that is too diverse, fast changing or massive for conventional technologies, skills and infrastructure to address efficiently. Big data is the collection, processing and availability of huge volumes of streaming data in real time. The use of big data— large sets of data that can be brought together and observed to analyse patterns and

---

[1] https://www.ap-institute.com/big-data…/how-is-big-data-used-in-practice-10-use-cases….
[2] www.forbes.com/sites/gilpress/2013/05/09/a-very-short-history-of-big-data/.

S. Misra (✉) · D. Sur
Centre for Consumer Studies, Indian Institute of Public Administration, I. P. Estate, Ring Road, New Delhi 110002, India
e-mail: drsureshmisra@gmail.com

D. Sur
e-mail: deepikasur.nch@gmail.com

© Springer Nature Singapore Pte Ltd. 2018
U. M. Munshi and N. Verma (eds.), *Data Science Landscape*,
Studies in Big Data 38, https://doi.org/10.1007/978-981-10-7515-5_9

make better decisions to enhance efficiency adds significant value. Analysing the input data sources improves the quality of the output by data being real time, dynamic and high frequency. Organizations and Government are starting to understand how powerful big data can be used to for delivering higher productivity and better value to consumers and citizens. This will enable better efficiencies and help provide better services.

Using and analyzing big data requires high-grade technical tools. Multiple, high volume and continuous streams of data require faster inputting and processing, with defined analytical tools. Conventional softwares cannot handle the high incoming volume and different data structure real time. In addition, tools for reporting the analysis also require to be sophisticated. Complicated analytics is used in decision-making, by minimizing risks, and discovering valuable insights, leading to informed decisions and not random outcomes. Big data allows for narrower segmentation of consumers to provide for customized and tailored products or services, leading to better customer service, operational efficiency, more effective marketing and revenue opportunities and other business benefits.

2 Big Data and the National Consumer Helpline[3]

In the era of globalization and competition, excellence in customer service is the most important tool for sustained business growth. Consumer complaints are part of the business life of any corporate/public entity. This is more so for service organizations. Quality service to the consumer and consumer satisfaction should be of prime concern of any service provider/seller. The right to be heard includes the right to be assured that consumers interest will receive due consideration at appropriate forum. A grievance is a documented manifestation of dissatisfaction of a consumer. Such dissatisfaction, if left unaddressed and unresolved, could endanger the lifeline of the company and erode its image. The objective should be to minimize instances of consumer complaints and grievances through proper service delivery and review mechanism and to ensure prompt redressal of consumer complaints and grievances. In order to meet the increasing legitimate expectations of consumers for better, faster and more effective service and better quality products the producer/service provider should constantly endeavour to improve its service delivery standards and capabilities.

The National Consumer Helpline (NCH) is a public service project of the Union Ministry of Consumer Affairs, Food and Public Distribution, operating from the Indian Institute of Public Administration, New Delhi, under the umbrella of the Centre for Consumer Studies (CCS). CCS is a dedicated centre and acts a "think tank" for the research and policy-related issues on consumer protection and consumer welfare. The centre keeps abreast of the long-term policies to position itself

[3]www.nationalconsumerhelpline.in

as a major contributor to the identification of issues and priorities as well as solutions to ensure better protection of consumers.[4]

The need for the Government to set up the National Consumer Helpline was essentially to protect consumers against exploitation. The Consumer Protection Act is a benevolent legislation to protect the interest of consumers. The Directive principles of State Policy of the Constitution of India require India to be a welfare state. It becomes the responsibility of the Government in a welfare state to facilitate its citizens to assert this right. Without Consumer welfare, India cannot be a welfare state. Thus, the need for consumer protection was realized. The advent of globalization and a market economy has expanded areas that need intervention on behalf of the Government to protect the interest of consumers. NCH is one of the key initiatives taken by the Government of India to make consumers aware of their consumer rights.

NCH receives calls from aggrieved consumers across the length and breadth of our vast country seeking answers to simple queries as well as complex problems. Calling the NCH toll-free number 1800-11-4000 empowers consumers by providing avenues for a solution to their consumer-related problem—could be a bank transaction, problems with household products like refrigerators or washing machines, or simply—how to write an right to information (RTI). *"This is in keeping with the NCH vision statement"* A Nation of awakened, empowered and responsible consumers and socially and legally responsible Corporations.

As the helpline deals primarily with complaints on a daily basis, NCH actively promotes the concept of best *practices* on complaint handling management within an organization and has conducted meetings and workshops with companies on the topic of World-Class Standards for Complaints Handling Management Systems. This underlines the *mission statement* of NCH—*"To provide telephonic advice, information and guidance to empower Indian consumers and persuade businesses to reorient their policy and management systems to address consumer concerns and grievances adopting world-class standards."* The National Consumer Helpline provides consumers the following:

(a) **Alternate dispute redressal system (ADR): information and guidance**:

Provide information to consumers on products, services, company addresses, ombudsman, regulators and consumer forums. Counsellors provide information as per the stage of the complaint—Tier 1, 2 or 3. Information is also provided for allied services like hallmark, ISI, RTI, PAN card, UIDAI, financial Inclusion programs of the Government, etc., and guide consumers on ways to get their grievance redressed.

(b) **Alternate dispute redressal system (ADR): convergence mode**

Under the convergence model, which is an alternate dispute redressal mechanism, NCH partners with companies who have a proactive approach to efficient consumer

---

[4]http://consumeraffairs.nic.in/forms/contentpage.aspx?lid=642.

complaint redressal. It shares with them the data of the complainants along with the complaints received at NCH related to their company, to facilitate free, fair and fast redressal through participative cooperation. Under the convergence process, companies are expected to resolve/redress and close the complaints by stating "a gist of the resolution" in the remarks column which is accessible to both—the consumer and NCH.

The complaints that are received at NCH are generally of an escalated nature, as most of the complainants have already gone through the first level—that of approaching the company. It is only when their complaints are not handled up to the complainants satisfaction that the complaints are lodged with NCH. It is precisely for this reason that NCH partners with various companies whereby these complaints are brought to the notice of a senior official of the company for a relook and offer a resolution, so as to dissuade consumers from moving to a consumer forum and also reducing litigation.

NCH has around 200 companies as partners in this effort, under the header Convergence@NCH. This is essentially an alternate dispute resolution method and consumer disputes are resolved cordially and mutually.

3 Partnering Agencies: Benefits

The benefits that a company gets in partnering with NCH are multifold as reflected here. Convergence is a win–win situation for both consumers as well as companies/ service providers by resolving consumer grievance amicably and expeditiously. It gives an opportunity for better corporate governance and social responsibility. Big data can be used to better understand customers and their behaviours and preferences. Companies are keen to expand their traditional data sets with social media data, browser logs as well as text analytics and sensor data to get a more complete picture of their customers.[5] The big objective, in many cases, is to create predictive models. Big data analytics also help machines and devices become smarter and more autonomous. Big data is also increasingly used to optimize business processes. Retailers are able to optimize their stock based on predictions generated from social media data, Web search trends and weather forecasts.[6]

- A proactive approach in resolving their customer grievances.
- Finding an amicable and expeditious resolution to the customers' problem, leading to a win–win situation for both—customers and organization, as there is a need for consumer disputes to be redressed.
- Gives an opportunity to the organization for an amicable settlement and retention of the customer, building customer loyalty.

---

[5]https://www.ijircce.com/upload/2016/june/55_A%20Study.pdf.

[6]http://www.datasciencecentral.com/profiles/blogs/the-awesome-ways-big-data-is-used-today-to-change-our-world.

- Ensures better corporate governance and social responsibility.
- Customer retention.

National Consumer Helpline also supports consumers by:

- Guiding consumers in finding solutions to problems related to products and services.
- Providing information related to companies and regulatory authorities.
- Facilitating consumers in filing complaints against defaulting service providers.
- Empowering consumers to use the consumer grievances redressal mechanisms.
- Educating consumers about their rights and responsibilities.

4 Big Data Handling at NCH

The National Consumer Helpline has been in operation for the last 10 years, resulting in a huge database of not only consumers and marketers but also of the nature and type of complaints and the pattern of redressal of these complaints. To large extent, it also explains how consumer sensitive and friendly is the business. Data is maintained in the following manner;

- Primary data is maintained on a month-wise basis.
- Data analysis is done on two fronts—routine and exceptional.
- Routine analysis of data is done on monthly basis. Monthly compilation of data is done to make state wise, sector wise, gender wise, company wise, and socio-economic profile of the consumers.
- Trend analysis on frequently occurring problems and enquiries is done on quarterly basis.
- Exceptional analysis is done for specific purposes.
- Data analytics is available to all stakeholders—in public domain.
- Database is confidential to ensure data privacy and security.

5 Data Management

The helpline follows stringent procedures for data management to ensure confidentiality, accuracy, authenticity while maintaining the transparency in final tangible outputs. All the data and its analysis reflected in the tables and figures are based on the data captured by counsellors through various modes and means indicated below. However, it may be highlighted here that the captured data is essentially primary in nature.

6 Primary Data Collection

Consumers calling NCH are advised telephonically by trained counsellors after capturing significant information from the consumer in the NCH software deployed. Information inputted includes name, age, city, state, e-mail and details of the complaint. This is further categorized to sector, category problem reported and, finally, the advice given by the counsellor.

Another area of data collection are the complaints that consumers log into the Website—www.nationalconsumerhelpline.in, where the consumer is required to provide significant information before lodging the complaint. This online data collection is another touch point where significant information is captured like name, age, city, state, e-mail and details of the complaint which is further categorized into sector, category, problem reported and the advice given by the counsellor.

7 Data Processing

This data collection is the backbone of the functioning of the National Consumer Helpline. This data collected undergoes processing to convert it into information. This is done through a series of steps. After the data is collected, it is then checked whether the mandatory fields are not left blank, and duplicity check is done. All blank and duplicate entries are not considered. Then, randomly gender, state, districts, etc., are checked through filters in excel. All Website complaints are also checked for errors and accuracy. Then, data is compiled and datasets are made. This forms the input data for processing. The data then goes through various filters and formulas in the computer, at the end of which output data is obtained. Output data can be in the form of charts, graphs, reports, etc. This output data is transmitted to the analyst, who then studies and interprets them and converts it into information. The information gathered from the analysis of the data provides the inputs and fodder for the major stakeholders involved in the business of handling consumer grievances as well as inputs for policy making. Table 1 depicts the fluctuations year wise of the first-level compilation of data of different sectors analysed at NCH, in descending order.

Calls received in products, telecom, banking and insurance sector have reduced in the year 2014–15, and there is a huge jump in the calls of e-commerce sector. There is a 54% increase in the year 2014–15 as Website complaints have been added.

Table 1 Sectoral analysis of data

Sector-wise calls received at NCH							
S. No.	Sectors	Apr 2013–Mar 2014		May 2014–Mar 2015		Apr 2015–Mar 2016	
		Calls in numbers	%	Calls in numbers	%	Calls in numbers	%
1	Products	23,252	23.10	30,322	21.60	36,401	21.09
2	Telecom	16,177	16.07	21,037	14.98	28,809	16.69
3	E-commerce	5204	5.17	15,168	10.80	28,331	16.42
4	Banking	5553	5.52	6826	4.86	8702	5.04
5	LPG/PNG	3330	3.31	5939	4.23	4417	2.56
6	Other sectors	47,144	46.83	61,116	43.53	65,918	38.20
	Total	100,660	100.00	140,408	100.00	172,578	100.00

8 Trend Analysis of Nature of Complaints

The analysis of data brings out the trends of complaints in each of the sectors. The frequently occurring problems reported in three major sectors are in Table 2.

Table 2 Trend analysis of nature of complaints

S. No.	Frequently occurring problems	2013–14 (%)	2014–15 (%)	2015–16 (%)
(a) Sector: products				
1	Service centre/dealer not entertaining	46	34	13
2	Delay in/not providing services	4	32	29
3	Same problem persists after repairs	15	14	11
4	Defective product—no repair/replacement	14	2	19
5	Dealer's misbehaviour	11	3	1
6	Non-delivery of product	2	3	2
7	Charging for repair under warranty	1	2	5
8	Other complaints	7	10	20
	Total	100	100	100
(b) Sector: telecom				
1	Unfair deductions	14	17	11
2	Inflated bills/overcharging	8	7	9
3	Activation of unsubscribed services	7	4	5
4	Delay in providing/activation of services	7	5	5
5	Wrong promises	5	3	5

(continued)

Table 2 (continued)

S. No.	Frequently occurring problems	2013–14 (%)	2014–15 (%)	2015–16 (%)
6	Unsatisfactory redressal	5	7	5
7	Broadband/internet not working/slow speed	4	8	10
8	Other complaints	50	49	50
	Total	100	100	100
(c) Sector: e-commerce				
1	Non-delivery of product	25	23	24
2	Delivery of defective product	12	18	18
3	Paid amount not refunded	10	14	8
4	Delivery of wrong product	7	10	8
5	Wrong promises	6	5	8
6	Deficiency in services	2	8	7
7	Other complaints	38	22	27
	Total	100	100	100

9 For Exceptional Analysis

- For specific and pilot studies, collecting "critical mass" of data is imperative. This means necessary or sufficient amount of data to have a significant effect or to achieve a result. This is required to be done for policy inputs, research papers and advocacies (Table 3).

Table 3 Infosource for reaching NCH

Top infosource of the NCH toll free number				
S. No.	Infosource	April 2013–March 2014 (%)	May 2014–Mar 2015 (%)	Apr 2015–Mar 2016 (%)
1	Internet (Primarily Google)	34.28	47.28	50.5
2	Word of mouth—reference	24.30	19.95	15.31
3	TV channels	15.91	10.56	5.37
4	Newspaper	7.62	5.60	3.74
5	Radio programs	2.78	1.34	6.47
6	Outdoor	0.95	0.99	0.46
7	Magazine	0.31	0.23	0.11
8	Events	0.05	0.05	0.68
9	Others	13.82	14.01	17.36
	Total	100.00	100.00	100.00

Table 4 Gender-wise analysis of data

Gender-wise analysis of calls			
	Apr 2013–Mar 2014 (%)	May 2014–Mar 2015 (%)	Apr 2015–Mar 2016 (%)
Male	91	92	92
Female	9	8	8

The above analysis is done to understand the source from where consumers have got to know about the toll-free number of the National Consumer Helpline. The top source is the Internet, followed by word of mouth—reference from an existing user of NCH. We also see a surge in the desired information being obtained from the Internet in the year 2014–15—up from 34 to 47% and a decline in the next two information sources. The other information sources for the toll-free number show no marked increase or decline (Table 4).

This is for a study done to analyse the gender-wise calls and complaints to the helpline. This is important for the decision maker to ascertain which media to publicize the helpline numbers and Website address. Further analysis on this data—state wise, district wise, etc., will help in narrowing down even further, to decide on the optimum answer.

This data may be explored on a different dimension as well—the operations of the helpline can be tweaked to encompass a specific objective—e.g., having more

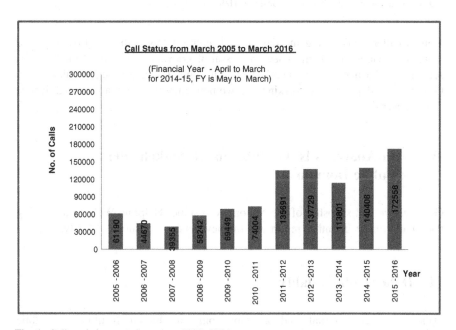

Fig. 1 Call statistics over the years: 2005–2016

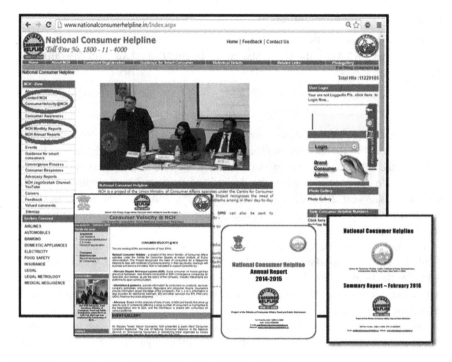

Fig. 2 A snapshot of the National Consumer Helpline Portal

women callers call during the day time, keeping a few dedicated lines only for women, to encourage them to seek redressal of grievances. It can also help social scientists and academicians to seek specific insights. It can also help regulators and Government departments working on women empowerment and social justice (Figs. 1 and 2).

10 Data Analytics is Available to All Stakeholders in Public Domain

Given above is a snapshot of the National Consumer Helpline Website displaying the annual report, monthly report and the newsletter—consumer velocity.

11 Data Confidentiality

Database is confidential and available only to Indian Institute of Public Administration, Department of Consumer Affairs, Government of India, and convergence partners. Convergence companies are those who voluntarily have partnered with the National Consumer Helpline to resolve their customer's grievance lodged with the helpline. These companies get access to only their complainants' data.

Hence, to summarize, consumer information captured from different channels, viz.—voice, SMS, e-mails, postal or snail mail are the major input data sources, and all these converged current and real-time data, with or without coupling with preceding data, can be studied to give a multidimensional perspective to decision makers and stake holders. This information is also used as inputs for the various training programmes conducted for improvement in counselling services by the helplines—both state and national as well as the Grahak Suvidha Kendras, all projects supported by the Department of Consumer Affairs, Government of India.

12 How Big Data Helps Consumers

Big data analysis helps in understanding the changing taste and preferences of consumers. This is very well brought out by studying the e-commerce market. Over the last five years, the share of e-commerce complaints has gone up, as shown in the chart along side. This reflects the growth of the e-commerce industry in the country.

Even in this sector, almost 70% complaints pertain to products bought online, and the remaining 30% caters to services. This reflects the shifting of preference of consumers from retail to online shopping (Fig. 3).

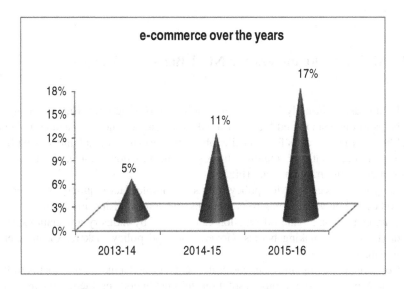

Fig. 3 Correlation between the market and consumer complaints

- Big data analytics reflects the pulse of consumer complaints in real time, as analysis of data helps all stakeholders to understand the trends in complaints in different sectors of the marketplace.
- In 2014, one of the largest e-commerce companies in India brought out a special festival sale for one day. There were too many glitches in the sale process, and NCH was inundated with calls and grievances from consumers. This gave real-time information of the market situation with regard to the sale. This real-time market information helped the company to rework on the details and processes, which led to a much more acceptable version of the sale the following year.
- Identify misselling and frauds taking place in specific sectors: The financial literacy amongst the consumers in India is fairly low. Even the otherwise literate consumers in India are financially illiterate. We observe this real time in the sales related to the insurance sector. The level of misselling in this sector is very high. The pie chart below depicts this problem. Consumers tend to believe the insurance agents word. They do not take out the time and make the effort to go through the fine print of the terms and conditions. So finally, when the verbally promised benefits do not come, the consumer is left high and dry.
- Understand and inform the consumer's viewpoint on product or service improvements for various companies in different industry segments, as a neutral organization.
- Data analysed of convergence companies—companies that have partnered with NCH helps in gauging the consumer satisfaction of the internal redressal mechanism of companies.

13 Major Stakeholders for NCH Big Data Analytics

- Data is used primarily for research—short term or long term. Big data analytics helps to understand and highlight trends in consumer enquiries and complaints.
- Short-term research is for a specific objective—to know complaints arising from a particular state or region, efficacy of information source, feedback for awareness programmes, etc. (Fig. 4).
- Long-term research is for policy inputs as sample data required by governing bodies to know the trends of complaints and issues arising frequently.
- Outcome of Data—the information can be used by industry associations, standards and law making bodies, Government and policy makers, social science researchers.
- Convergence companies can check the scope of consumer requirements to find amicable and expeditious resolution to customers' problem, leading to a win-win situation for both—customers and corporates and give an opportunity

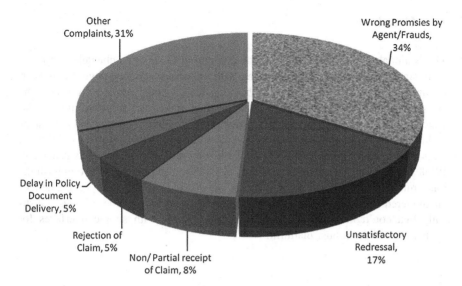

Fig. 4 Complaint characteristics

to the organizations for an amicable settlement and retention of the customer, building consumer loyalty.

14   Major Challenges in NCH Big Data

- The data generated is huge, and requires enormous resources for standardization, interpretation, communication and presentation.
- There is not much response for the information generated from the data analysed from companies—manufacturers and service providers.
- Government response to data is limited as the departments are not able to access and customize the information as per their requirements. No designated nodal officers are there in various organizations to use the data, hence not used much in policy analysis/formulation.
- Data is in abundance; however, NCH does not have an intelligent analytics system for counsellors to share insights from the data analytics to be able to better respond to callers, real time.
- The primary constraint is lack of trained manpower to analyse and use big data tools.

15 Conclusion

The National Consumer Helpline while continuing to serve its clientele and make them smart consumers is also grappling with the issues of data deluge. In an effort to make them smart consumers, the helpline fosters empowerment by giving the consumers guidance, educating them about their rights and responsibilities, grievance redressal system, providing information related to companies and regulatory authorities and also guiding them to provide solutions to their problems. All these transactions leave digital trails and also in the long run, colossal data gets generated. While the helpline tries to indulge in data analysis to decipher possible trends, in the long run with more and more data on one hand and the kind of challenges encountered as reflected above, there is a need for addressing these issues holistically that could facilitate in tackling complex issues with simple interfaces for furtherance of the robust platform.

Big Data in the Context of Smart Cities: Exploring Urban Planning and Governance

Sushma Yadav and Gadadhara Mohapatra

1 Introduction: Understanding Big Data

There is a long history of government, business, science and citizens producing and utilizing data to monitor, regulate, profit from and make sense of the world.[1] Traditionally, data generation has been perceived as time-consuming and it is a costly affair to generate, interpret, and analyze. Moreover, it provided a static notion of social phenomena. However, over the years, the production and nature of data is being transformed through a set of what Christensen (1997) terms 'disruptive innovation' that challenge the status quo as to how data are produced, managed, analyzed, stored and utilized. Rather than being considered scarce and limited in access, data generation is increasingly becoming a deluge; a wide, deep torrent of timely, varied, resolute, and relational data that are relatively low in cost and outside of business, increasingly open and accessible. The dynamic nature of data generation is aptly defined by Kitchin (2014) as 'Data Revolution.' He indicated that a data revolution is underway, which is reshaping the modes production of knowledge, conduct of business and government. It also raises many ethical questions concerning privacy, security, surveillance, profiling, social sorting, and intellectual property rights. This data revolution is primarily based on the latest wave of information tools and communication technologies (ICTs) and the Internet of things (IoT).[2]

---

[1]Kitchin (2014: *preface*: *xv*).
[2]Ibid.: xv.

S. Yadav (✉) · G. Mohapatra
Indian Institute of Public Administration, I.P. Estate, Ring Road,
New Delhi 110002, India
e-mail: sushma.iipa@gmail.com

G. Mohapatra
e-mail: mohapatra.iipa2014@gmail.com

© Springer Nature Singapore Pte Ltd. 2018
U. M. Munshi and N. Verma (eds.), *Data Science Landscape*,
Studies in Big Data 38, https://doi.org/10.1007/978-981-10-7515-5_10

The data revolution is not only confined to the computational sciences. Although the initial journey has been in the commercial world, it also relates to data about patterns of human interaction. The United Nations Secretary-General's Independent Expert Advisory Group Report, *A World That Counts: Mobilising the Data Revolution for Sustainable Development* (2014),[3] lays out the potential policy implications for social sciences. The World Economic Forum, noted in its 2014 Global Technology Reports that data is an asset. It promises a future where decisions about business, life, and society will be taken purely based on data.

2 Big Data Characterization

There is no universal definition of big data. As pointed out by, Mayer-Schönberger and Cukier (2013, pp. 7),[4] 'the real revolution is not in the machines that calculate data but in data itself and how we use it.' According to Gartner: 'Big Data is high volume, high velocity and high variety information assets that demand cost-effective, innovative forms of information processing for enhanced insight and decision making.'[5]

Francis Diebold (2012)[6] traces the genesis of the term 'big data' to the mid-1990s, first used by John Mashey, retired former Chief Scientist at Silicon Graphics, to refer to handling and analysis of massive datasets. Initially, the term had little traction. Till 2008, very few people were using the term 'big data,' either in the academia or industry. Five years later, it had become a buzzword, commonly used in business circles, academia, and popular media. Diebold (2012) opined that big data consist of three things such as: the term (firmly entrenched), the phenomenon (continuing unabated), and an emerging discipline.[7]

The term big data is commonly associated with the 3Vs: volume, velocity, and variety.[8] The fourth 'V' in big data deployment as articulated by the data analytics

---

[3]United Nations Data Revolution Report (2014). A World That Counts: Mobilising the Data Revolution for Sustainable Development. Available at http://www.undatarevolution.org/report/.

[4]Mayer-Schonberger, V. and Cukier, K. (2013). Big Data: A Revolution That Will Change How We Live, Work and Think. John Murray, London.

[5]Gartner (2013). Gartner IT Glossary Big Data. Available at: www.gartner.com/it-glossary/big-data.

[6]Diebold, F. (2012) 'A personal perspective on the origin(s) and development of 'big data': the phenomenon, the term and the discipline', Available at: http://www.ssc.upenn.edu/∼fdiebold/papers/paper112/Diebold_Big_Data.pdf.

[7]Diebold (2012:5).

[8]Laney, D. (2001). 3D data management: controlling data volume, variety and velocity, META Group File 949. Available at: http://blogs.gartner.com/doug-laney/files/2012/01/ad949-3D-Data-Management-Controlling-Data-Volume-Velocity-and-Variety.pdf.

is the 'Value.'[9] Although solving the emerging challenges around the 3Vs is the focal point of many big data solutions; however, the realization of the fourth V-value is the key. Beyond the 3Vs, the emerging literature characterizes big data being exhaustive, fine-grained in resolution, relational in nature, flexible, and scalable.[10]

Sociologists define big data as a step toward change in scope and scale in relation to a particular phenomenon[11] such as urban governance and for this chapter smart cities. Generally, big data is understood as being born digital[12] as it is generated by the use of digital technologies; it consists largely of data which is distributed and therefore requires distributed analytical techniques.[13] Moreover, it often comprises data which is remotely collected and therefore either observed, namely, as a product/result of people's use of technology even though the production of data may not be the aim of using a given technology, or inferred, namely, information that is called out of existing data sources through big data analytics.[14] The distributed nature of big data offers complex analytical challenges. Thus, it requires new configurations of computing tools/technology and human collaboration to deal with it in the initial state.

The main characteristics of big data could be summed up as (i) It is observed or inferred (traditional digital data sources) but may also be volunteered (new sources, e.g., social media, crowdsourcing feedback); (ii) It is born digital: created in digital form; (iii) It is datafied: machine readable, with the potential to link, merge, and analyze across sources; (iv) It has the potential to bring out new issues/questions through merging/linking/distributed analysis; (v) In its original state, it presents significant storage and management challenges; (vi) In urban contexts, it requires collaboration between city authorities, experts, and data scientists.[15] Recently, Kitchin (2014)[16] has elaborated seven key characteristics of big data such as volume, velocity, variety, exhaustivity, resolution/indexicality, relationality, and flexibility/scalability that distinguish them from small data.

---

[9]Garasimou, Vicke (March 2016). Big Data and the 3Vs: What is the fourth 'V" and what are the implications for not embracing it? Available at: https://www.thinkbiganalytics.com/2016/03/29/big-data-3vs-fourth-v-implications-not-embracing/.

[10]Boyd and Crawford 2012; Dodge and Kitchin 2005; Marz and Warren 2012; Mayer-Schonberger and Cukier 2013, cited in Kitchin 2014, p. 68.

[11]Schroeder, R. (2014). Big data: Towards a more scientific social science and humanities? In: Graham M, Dutton WH (eds) Society and the internet: how networks of information and communication are changing our lives. Oxford University Press, Oxford, pp. 164–176.

[12]Borgman, C. (2014). Big data, little data and beyond. MIT Press, Cambridge.

[13]Taylor L, Schroeder R (2014). Is bigger better? The emergence of big data as a tool for international development policy. Geo Journal. https://doi.org/10.1007/s10708-014-9603-5.

[14]Hildebrandt, M. (2013) Slaves to big data. Or are we? IDP 17:27–44.

[15]Hildebrandt 2013 and Taylor et al. 2014, cited in Linnet Taylor and Christine Richter, Chapter 9; Big Data and Urban Governance, p. 177.

[16]Kitchin (2014:78).

The next section provides an explanation of the term smart city and describes the main components of smart infrastructure.

3 Smart City Concept

One can hardly find a universally accepted definition of a smart city. In 2014, the International Telecommunication Union report while analysing over 100 definitions related to smart cities, derived a broad definition of it such as: 'A smart sustainable city is an innovative city that uses ICTs and other means to improve quality of life, efficiency of urban operation, and services and competitiveness, while ensuring that it meets the needs of present and future generations with respect to economic, social, and environmental aspects.'[17] Further several efforts are made by the social scientists and data analytics to compose a set of some key performance indicators for the smart cities.

The United Nations inter-agency group has taken the lead in evolving a set of key performance indicators (PIs) with the aim to covert these into a global smart sustainable cities index.[18] A closer look into the meaning and definitions of smart city reveals that these definitions emphasize on various components of the term. The smart city concept offers different opportunities while posing different challenges for different countries. However, there is an urgent need for cities in developing countries to provide adequate urban infrastructure and the basic services to meet the increasing demands of urbanization. Smart infrastructure application could provide a road map for cities to achieve leapfrogging in technology.[19]

The various components of smart city are smart economy, smart people, smart governance, smart workforce, smart mobility, smart environment, smart spaces, and smart living. These combined components try to raise the economic competitiveness, social and human capital, participation, transparent governance and use of ICT, natural resources, and quality of life. Figure 1 shows smart city components.

The Smart Cities Mission is a new and innovative initiative of Government of India. In the approach to Smart Cities Mission, the objective has been to promote cities that provide core infrastructure facility and offer a decent quality of life to its

---

[17]International Telecommunication Union (2014). Smart sustainable cities: An analysis of definitions, Focus Group Technical Report, cited in UN ECOSOC Report (2016). Smart Cities and Infrastructure, p. 3, available at: http://unctad.org/meetings/en/SessionalDocuments/ecn162016d2_en.pdf.

[18]D Carriero, 2015, United smart cities: Towards smarter and more sustainable cities, presented at the CSTD intersessional panel; and B Jamoussi, 2015, Shaping tomorrow's smart sustainable cities today, presented at the CSTD intersessional panel.

[19]Deloitte (2014). Africa is ready to leapfrog the competition through smart cities technology, available at http://www2.deloitte.com/content/dam/Deloitte/za/Documents/publicsector/ZA_SmartCities_12052014.

SMART ECONOMY (Competitiveness)	SMART PEOPLE (Social and Human Capital)	SMART GOVERNANCE (Participation)

▪ Innovative spirit
▪ Entrepreneurship
▪ Economic image &
  trademarks
▪ Productivity
▪ Flexibility of labour market
▪ International embeddedness
▪ Ability to transform
	▪ Level of qualification
	▪ Affinity to lifelong learning
	▪ Flexibility
	▪ Social and ethnic plurality
	▪ Creativity
	▪ Cosmopolitanism/open-
	  mindedness
	▪ Participation in public life
		▪ Participation in decision
		  making
		▪ Public and Social
		  Service Delivery
		▪ Transparent Governance
		▪ Political strategy and
		  perspective

SMART MOBILITY (Transport and ICT)	SMART ENVIRONMENT (Natural resource)	SMART LIVING (Quality of life)

▪ Local accessibility
▪ (Inter-) National
  accessibility
▪ Availability of ICT-
  infrastructure
▪ Sustainable, innovative &
  safe transport system
	▪ Attractivity of natural
	  condition
	▪ Pollution
	▪ Environmental protection
	▪ Sustainable resource
	  management
		▪ Cultural facilities
		▪ Health conditions
		▪ Individual safety
		▪ Housing quality
		▪ Education facilities
		▪ Touristic attractivity
		▪ Social cohesion

Fig. 1 Smart city components (Skaily, Iryana. Smart City: City of Future. UNDP/MGSDP. Available at: https://www.slideshare.net/olenaursu/smart-city-27072012-engl)

citizens/inhabitants/residents, a clean and sustainable environment and use of smart solutions.[20]

An illustrative list of 'Smart Solutions' includes: (a) e-governance and citizen Services, (b) waste management, (c) water management, (d) energy management, (e) urban mobility, and others (see Fig. 2).

The next section provides a descriptive account of the significance of big data in the urban context.

4 Big Data in the Urban Context

The rise of the smart cities and smart citizens' concepts are often linked to the rise of big data. The smart city is an achievable utopia, an ideal-type urban constellation that addresses various aspects of urban life (transport, the environment, quality of life). As an ideal-type, smart city envisions about the future of the city, how a real smart city should look like. A group of scholars have viewed that the concept of smart city, in its vision and reality is a 'datafied' city[21] in the sense that it

---

[20]MoUD (2015: 5–6).

[21]Mayer-Schönberger, V., & Cukier, K. (2013). Big Data: A revolution that will transform how we live, work, and think. John Murray, London.

WASTE MANAGEMENT	ENERGY MANAGEMENT	OTHERS
1. Waste to Energy & Fuel 2. Waste to Compost 3. Waste Water to be Treated 4. Recycling and Reduction of C&D Waste	13. Smart Meters& Management 14. Renewable Sources of Energy 15. Energy Efficient & Green Buildings	19. Tele-Medicine & Tele Education 20. Incubation/Trade Facilitation Centers 21. Skill Development Centers
E-GOVERNANCE & CITIZEN SERVICE	WATER MANAGEMENT	URBAN MOBILITY
5. Public Information, Grievance Redressal 6. Electronic Service Delivery 7. Citizen Engagement 8. Citizens-City's Eyes and Ears 9. Video Crime Monitoring	10 Smart Meters & Management 11. Leakage Identification, Preventive Maint. 12. Water Quality Monitoring	16. Smart Parking 17. Intelligent Traffic Management 18. Integrated Multi-Modal Transport

Fig. 2 Smart solution (Smart Cities Mission Guidelines, Ministry of Urban Development, GoI, 2015, p. 6)

simultaneously produces and consumes streams of digital data, and the unprecedented amount of data generated with complexity and size.[22]

Big data requires and also allows for new forms of urban data analytics using data from a variety of sources such as sensor networks, electronic feedback systems, or social media platforms. Its potentials for urban design, administration, planning, business, and environmental management are manifold. Big data generated through pervasive sensing from Internet of things (IoT) from mobile phones, smart meters, and radio-frequency identification (RFID) tags could be further help in analyzing the resources and energy used in a city and how traffic and people are moving around the urban space.[23] The big data emanating from digital devices has the potential to be applied in planning for more efficient and sustainable energy use, traffic management, and transportation. The flow of data reflects the use of space, often produced and used by city service providers. These, if merged or linked, can provide a dynamic spatial image of how various city functions such as law enforcement and service delivery provision are working. Further, this data can be used for predicting the changing profile of crimes in a city and its prevention, solution, and emergency planning. Finally, crowdsourced data from social media or online communication applications such as feedback applications (apps) for

---

[22]Townsend, AM. (2013). Smart cities: big data, civic hackers, and the quest for a new utopia. W. W. Norton & Company, Cambridge.

[23]Pentland, A. (2011). Society's nervous system: building effective government, energy, and public health systems. Pervasive Mob Compute 7(6):643–659.

identifying gaps in service provision can provide authorities with a real-time picture of how the city and its residents are interacting, and where needs and gaps are signaled.[24]

The concept of big data is often controversial in terms of the way it is collected, used, disseminated and its societal, ecological, and scientific implications at large. Second, big data deals with the fast changing socio-technical arrangements which possess challenge in the established modes of thought and practice for scientists, citizens, and policymakers.[25]

According to Bauman et al. (2014),[26] digitalization in the public sector is sparking new constellations of professionals and 'hybridizing private and public actors.' While use of big data and data analytics can help in improving urban planning, governance and management; future research on big data and urban governance has to transcend beyond a mere techno-solutionist discourse, putting empirical, analytical, and theoretical emphasis onto a critical socio-technical understanding of the nature of new social and technical actor constellations and their related power dynamics.[27]

The next section looks at the key role and the potential use of big data in the promotion of smart cities.

5   Role of Big Data in Smart Cities

Studies highlight that globalization as a process is mutually shaping and being shaped by the geographies of urban governance. It is also shaping the dominant discourses that explain the way societies, cities, and the global community is organized.[28] The shift in the understanding from government to urban governance implies that there is growing involvement of private sector, citizens, and grassroots organizations in navigating urban affairs. In view of this, there is the necessity of more participatory approaches and new institutional arrangements like public–private partnerships.[29] This shift in understanding is perhaps rooted in the dominant discourses such as neoliberal capitalism, neo-institutionalism, and neorealism to the latest discourse on sustainable development, the green economy and inclusive

---

[24]Taylor, L. and Christine Richter (2015). Big Data and Unban Governance (Chapter 9, pp.175–191). In Joyeet Gupta, et al. (2015) Geographies of Urban Governance: Advanced Theories, Methods, and Practices. Springer International Publishing, Switzerland.

[25]Taylor and Richter (2015:177).

[26]Bauman et al. (2014: 126).

[27]Boyd D, Crawford K (2012). Critical questions for big data. Inf Commun Soc 15(5):662–679 Crampton JW, Roberts SM, Poorthuis A (2014) The new political economy of geographical intelligence. Ann Assoc Am Geogr 104(1):196–214.

[28]Gupta, Joyeeta et al. (2015:5). Geographies of Urban Governance: Advanced Theories, Methods, and Practices. Springer International Publishing, Switzerland.

[29]For further details on the topic Joyeeta Gupta et al. 2015, pp. 6–11.

development. This provides a conceptual frame within which new concepts of city life evolves, in contemporary policy debates we do talk about the city of the future such as: global city, ordinary city, the just city, the sustainable city, the smart city, the inclusive city, the ludic city.[30]

The discourse of the just city focuses on norms, values, and sociopolitical and economic rights. Just city is all about the city and its residents and it primarily focuses on promoting the modern values of democracy, diversity, and inclusiveness. It promotes the rights to the city and human flourishing and welfare.[31]

The concept of smart city focuses on data and efficiency. The smart city concept refers to cities whose governance increasingly depends on data generating from digital technologies on various aspects of public service management such as travel, communication, energy uses, water uses, waste flows, health care, suggestions, and complaints from online data collection systems in order to both engage with residents and to provide the services needed by residents in a more effective, efficient, inclusive, and sustainable manner.[32]

Finally, the sustainable city discourse focuses on socioeconomic and ecological issues. The sustainable city[33] integrates socioeconomic and ecological perspectives based on the principles of equity, the no-harm principle and sustainability based on procedural and interspecies equity.[34] Sustainable cities provide an environment which ensures livelihood opportunities for all the citizens; a safe, secure environment for people with minimal resource use and pollution of ecosystems; and the freedom as well as opportunity to participate in politics.[35]

6 Big Data Initiative of Government of India for Smart Cities

The Government of India has launched a Big Data initiative. The Digital India program is a flagship program of the Government of India. It is based on a vision to transform India into a digitally empowered society and knowledge economy. It aims at ensuring that the government services are made available to citizens electronically by reducing physical movement and paperwork. The National Informatics Centre (NIC) has created the Open Government Data portal, data.gov.in. In the year 2014,

---

[30]Ibid. p. 6.

[31]Uitermark J (2011). An actually existing just city? The fight for the right to the city in Amsterdam. In: Brenner N, Marcuse P, Mayer M (eds) Cities for people, not for profit: critical urban theory and the right to the city. Routledge, Abingdon/New York, pp. 197–214.

[32]Hollands RG (2008). Will the real smart city please stand up? Intelligent, progressive or entrepreneurial? City 12(3):303–320.

[33]Satterthwaite D (ed) (1999). The Earthscan reader in sustainable cities. Earthscan, London.

[34]Haughton, G. (1999) Environmental justice and the sustainable city. J Plan Educ Res 18(3): 233–243.

[35]Kitchin (2014:6).

the Indian Government had announced building 100 smart cities across the country in light of the shift toward urban transformation due to massive influx of migrants from villages.

While several Smart Solutions have been proposed by Cities as part of their Smart City proposals, this sector in India is still in nascent stage and there is vast scope for creativity and innovation.[36] In view of this, the Ministry of Urban Development (MoUD), Government of India has launched a scheme for 'Promoting Innovative Smart Solutions under Smart City Mission, AMRUT and Swachh Bharat Mission' with the twin objective of providing a flip to the development of innovative solutions that can impact the needs of cities under the Mission, and making available a large body of pilot tested and proven solutions that can be adopted by cities as per their needs. The scheme aims to provide funding support for industry-sponsored, outcome-oriented projects focused around the needs of Urban-Local Bodies (ULBs as per Census 1991, classified into four major categories of: Municipal corporation (Nagar Nigam), Municipality (Nagar Parishad), Town area committee and Notified area committee) under Smart City Mission, AMRUT, and also the Swachh Bharat Mission.[37]

However, the Mission lacks clarity and has faced several challenges regarding use, storage, and ownership of such data in the cities, the actors involved and their accountability, concerns around data security, privacy, and need for a suitable regulatory framework.[38]

For smart cities, smart governance is necessitated so as to factor transparency in the urban context. As a precursor, the launch of Unique Identification Authority of India (UIDAI) as a statutory authority was established under the provisions of the Aadhaar (Targeted Delivery of Financial and Other Subsidies, Benefits and Services) Act, 2016 (Aadhaar Act 2016) on July 12, 2016 by the Government of India, under the Ministry of Electronics and Information Technology (MeitY). The sole purpose of UIDAI was to create a provision for issuing a Unique Identification numbers (UID), named as 'Aadhaar,' to all residents of India that brings in (a) robustness in system to eliminate duplicate and fake identities and (b) facilitate verification and authentication in an easy, cost-effective way.[39] Besides, the data availability initiative being facilitated by NIC[40] and buttressed by the launch of the National Data Sharing and Accessibility Policy[41] adds to data discoverability for

---

[36]Innovations for Smart Solutions, March 20, 2017. Available at: https://smartnet.niua.org/sites/default/files/resources/Innovation_for_Smart_Solutions_-_final_20%20mar17.pdf.

[37]Ibid.: p. 3.

[38]Big Data and Governance in India: A Discussion on the Role of Big Data in Urban Governance in India with a focus on Digital India, UID Scheme and Smart Cities Mission. Available at: https://cis-india.org/internet-governance/blog/background-note-big-data.

[39]Unique Identification Authority of India, Govt. of India. Available at: https://uidai.gov.in.

[40]National Informatics Centre (NIC), Govt. of India. Available at: http://www.nic.in/.

[41]Ministry of Science and Technology (2012). National Data Sharing and Accessibility Policy (NDSAP) 2012. Gazette of India. March 17. http://data.gov.in/sites/default/files/NDSAP.pdf.

better decision making and effective governance. It offers space for sharing information on available data, which could be shared with civil society at large with developmental objectives. The scope of the policy is limited to data owned by the agencies, departments/Ministries, and other entities under the Government of India and forms a statement of the Government of India of its commitment to transparency and efficiency in governance.

However, there are lot of threads needed to be woven before we arrive at a robust framework for smart cities that is truly smart in nature and flavor and that can holistically reflect in overcoming urban crisis (e.g.; transport, energy issues, water and sanitation, and the like) that are otherwise looming large.

7 Conclusion

Smart city and big data are two nascent concepts. If developed into smart city applications, big data could further help to reach sustainability, better resilience, effective urban governance, improved quality of life, and intelligent management of smart city resources.[42] The chapter has sought to explore both the concepts of big data and smart city and has in the process described the changing and expected components of smart city in terms of urban infrastructure and urban governance. The exploration reveals that there is a huge scope of big data in the development of sustainable smart cities, analyzing the components smartly, and providing smart solutions to the problems of urban governance in India. Yet certain issues and concerns need to be addressed to achieve better utilization of smart technology. Nonetheless, the exploration brings out the potentials and opportunities that big data offers in urban planning and development of smart cities. It can be concluded from the above exploration that the smart city mission needs more clarity and has faced several challenges regarding use, storage, and ownership of such data in the cities, the actors involved and their accountability, concerns around data security, privacy, and need for a suitable regulatory framework. In a nutshell, smart thinking and smarter application of technology is the need of the time.

---

[42]Nuaimi, E Al. et al.(2015). Applications of Big Data to Smart Cities. Journal of Internet Services and Applications, December Issue, pp. 3–15.

Crowd Sourcing for Municipal Governance

K. K. Pandey

1 The Background

Recent 10–15 years have witnessed growing attention towards Internet-based data collection and placement for wider dissemination. The term "crowd sourcing" was coined in 2005 by Jeff Howe and Mark Robinson, editors at *Wired*, to describe how businesses were using the Internet to "outsource work to the crowd", which quickly led to the *portmanteau* (multipurpose) "crowd sourcing".[1] Enrique Estellés-Arolas and Fernando González Ladrón-de-Guevara, researchers at the Technical University of Valencia, developed a new integrated definition on the basis of a survey of literature on the subject covering 40 definitions reaffirmed achievement of multipurpose, multi-stakeholder tasks through crowd sourcing[2] (Box 1).

> **Box 1**
> **Definition of Crowd Sourcing**
> Crowd sourcing is a type of participative online activity in which an individual, an institution, a non-profit organization or company proposes to a group of individuals of varying knowledge, heterogeneity and number via a flexible open call, the voluntary undertaking of a task. The undertaking of the task, of variable complexity and modularity, and in which the crowd should

---

[1]Safire, William (February 5, 2009). "On Language". New York Times Magazine. Retrieved May 19, 2013.
[2]Estellés-Arolas, Enrique; González-Ladrón-de-Guevara, Fernando (2012), "Towards an Integrated Crowd sourcing Definition" (PDF), *Journal of Information Science*, **38** (2): 189–200, https://doi.org/10.1177/0165551512437638.

K. K. Pandey (✉)
Centre for Urban Studies, Indian Institute of Public Administration, New Delhi, India
e-mail: kkpandey9236@gmail.com; kpandey_2000@hotmail.com

© Springer Nature Singapore Pte Ltd. 2018   151
U. M. Munshi and N. Verma (eds.), *Data Science Landscape*,
Studies in Big Data 38, https://doi.org/10.1007/978-981-10-7515-5_11

participate, bringing their work, money, knowledge **[and/or]** experience, always entails mutual benefit. The user will receive the satisfaction of a given type of need, be it economic, social recognition, self-esteem or the development of individual skills, while the Crowd Source will obtain and use to their advantage that which the user has brought to the venture, whose form will depend on the type of activity undertaken.

As other stakeholders, city governments also gradually adopted crowd sourcing across the countries. The process is still followed and forward-looking cities are presenting models for wider adaptation. This paper examines use of crowd sourcing in the urban/municipal governance in India and tries to shape up a follow-up agenda for wider information and dissemination.

2 Urban Sector in India and Crowd Sourcing

Government of India (GoI) has a well-developed website which has efficient links with its subordinate offices and concerned agencies. Recently, Government of India has merged its two urban ministries as Ministry of Housing and Urban Affairs (MoHUA).[3] However, urban development is state (provincial—hereafter used as state) subject in the federal structure of India. Accordingly, the strategy and follow-up on crowd sourcing are determined by respective state governments. GoI also provides support, guidance, handholding on the subject through respective states. The successive programmes and schemes of GoI have given due cognizance to crowd sourcing in the urban sector.

The information of urban sector websites giving complete details is not available at a single source. GoI web directory shows only 227 cities and other urban institutions with their exclusive websites which are highly underreported.[4] Most of the 194 municipal corporations have their own website[5] as a result of focus of successive programmes and policies. The number of other cities and institutions with their exclusive website is significantly more than the no of corporations. The pattern of application varies significantly from state to state. Some states such as Jharkhand and Madhya Pradesh have link in the state website indicating city website or specific function such as payment/assessment of property tax, birth and death registration. Data on city websites is fairly ad hoc. It is not regularly updated in a real-time reporting. The construction of city website varies significantly from

---

[3]http://moud.gov.in.

[4]goidirectory.nic.in (e-governance has been the part of reform agenda of GoI since 2005 as part of national missions).

[5]http://infoelections.com/infoelection/index.php/indian-politics-53015/about-indian-politics-43710/5990-municipal-corporations-in-india.html.

Table 1 City government websites in select states

State	Delhi NCT	Karnataka	Kerala	Tamil Nadu	West Bengal
City government website	4 (4)	190 (220)	56 (59)	125 (721)	122 (129)

Figures in the brackets relate to city governments as per Administration of Urban Development and Urban Service Delivery, IIPA, KK Pandey

state to state as compared to number of statutory towns (city governments) as per census 2011 (Table 1).

Some states are in a process to activate the city-specific websites. Madhya Pradesh, for instance, shows links for 16 websites for municipalities whereas 32 are under activation/construction. Tamil Nadu shows links for city websites. As may be seen from Table 1, the process is taken up by states at different levels of development.

Therefore, state-level websites for nodal agencies (Directorate of Municipalities in most cases) and central government website need to be linked with municipal websites. MoHUA website in this regard needs special attention to guide urban sector stakeholders to access crowd sourcing in an user-friendly manner. It will provide a proper assessment urban sector crowd sourcing and feedback for policies, grievances and monitoring.

The scope in this connection is fairly wide (Table 2). At the same time, mobile-based applications are available to all the 8000 + urban centres which provide a basis for expansion of web-based applications to link common man with municipal database through crowd sourcing. Many forward-looking cities have initiated a network of apps and Whatsapp groups to engage community with the municipal system and initiatives.[6]

As per recent estimates, 24% Indian population accounting for 332 million persons have access to Internet. It is equally revealing to note that out of these, 323 million accounting for 97% have access through mobile services.[7] Thus, mobile users form the largest pool of potential consumers of crowd sourcing.

As may be seen from Table 3, the access to mobile is increasing the scope of crowd sourcing across the country. Urban India with an estimated population of 444 million already has 269 million persons (61%) using the Internet. Rural India, with an estimated population of 906 million as per 2011 census, has only 163 million (17%) Internet users.[8] It is further important to note that the access to mobile to bottom quintile in urban area is as high as 77%, whereas they significantly lack access to basic services such as tap water (only 18%), paved roads, sewage, street lighting, solid waste collection etc. Therefore, crowd sourcing (Internet/mobile connectivity) can significantly promote inclusive municipal governance at grass-roots level.

---

[6]The city of Bengaluru and Pune have several apps and Whatsapp groups to monitor and supply waste management services in a citizen-centric manner.

[7]Times of India, 18 September 2017.

[8]Mint 14 August 2017.

Table 2 Websites for urban agencies

Urban institutions/agencies	No's	Exclusive websites	% of No's
Urban local bodies	4041	259	5.75%

Table 3 Mobile connectivity in urban areas

Item	Urban areas (%)
All population with Internet1	60–62
Bottom quintile with mobile2	77
Bottom quintile with tap water2	18

Source 1. Live Mint 5 December 2016: 2. Live Mint 16 March 2017

3 City-Scale Crowd Sourcing

Urban sector institutions including city governments in India gradually realized the utility of crowd sourcing and initiated steps accordingly. The state of Tamil Nadu is one of the early reformers who adopted crowd sourcing and engaged cities through hackathon (Box 1). This has encouraged cities in the state to develop websites and start portals, apps and links for various types of information sharing, feedback and grievance redressal. Subsequently, Jawaharlal Nehru National Urban Renewal Mission (2005–2012) designed e-governance as an integral part of municipal reform agenda. The current scheme and programmes, namely Smart City Mission, Atal Mission for Rejuvenation and Urban Transformation (AMRUT) and Urban Transformation, Swachh Bharat Mission, give due emphasis on crowd sourcing.

Box 2

Hackathon for Urban Governance

Government of Tamil Nadu has the credit to have one of the initial efforts of municipal sensitization for crowd sourcing in the state in December 2013. The Commissionerate of Municipal Administration (CMA) the nodal arm of provincial government to engage cites and towns in the state in association with Anna University in Chennai and global technology consultancy, organized a two-day hackathon on 13–14 December which witnessed active participation from 172 technology professionals and students, who came together to build web and mobile applications to help improve lives of citizens in municipalities in Tamil Nadu.

These professionals developed 25 applications on sanitation, civic services, education, health, birth and death registrations among other areas.

Participants designed and shared ideas based on actual data collected from 135 municipalities in Tamil Nadu for this event. The hackathon also deliberated on the need for good governance to bridge society's ideas with social media, mobile computing, visualization using analytics and cloud + app usage. Care was taken to cover basic idea, usability and creativity, technology design and efficiency and quality. CMA subsequently implemented applications municipal level to improve urban governance by boosting engagement between municipal employees and citizens with real-time updates and reporting of day-to-day updates.

These efforts duly facilitated wider replication and due cognizance by state government, and World Bank supported Tamil Nadu Urban Development Programme (TNUDP) to implement a wider agenda for urban governance in the state.

Source: http://cma.tn.gov.in/cma/en-in/Pages/Hackathon.aspx and discussions with urban sector functionaries in the state.

4 Emerging City Models

Use of crowd sourcing involves a range of actions at city level to apply crowd sourcing for improved urban governance. In this regard, it is attempted in the following analysis to share models of municipal efforts to link crowd sourcing with the needs of a common man. As an illustration, we have covered three important sets of functions which are used by specific city as a model for wider understanding. These include (i) municipal accounting and revenue administration, (ii) e-governance known as e-service centres as applied in Ahmedabad Municipal Corporation (AMC) and (iii) management of municipal roads, solid waste and assets as applied by Bruhat Bengaluru Mahanagara Palike (BBMP).

5 Accounting and Revenue Administration

All the mega cities in India have a history to initiate crowd sourcing through wide area network (WAN) and local area network (LAN) for municipal system of revenue and services. Ahmedabad is one of the initial city starting WAN from the year 1998 for city-scale application of online entries and updating of data on revenue receipts and expenditure along with assets. It has improved transparency, efficiency and accountability in the financial management. At the same time, online collection of taxes was also included to have remarkable increase in the collection of revenue. It has provided access to a large number of citizen.

The process covers a range of actions as a prerequisite to generate data for subsequent placement in the crowd sourcing. As may be seen in Table 4, out of box thinking, forward-looking approach and handholding by provincial and national government have paved way to link crowd sourcing with the ongoing municipal reforms in the delivery of respective municipal functions, double-entry accounting,

Table 4 Crowd sourcing for municipal accounting and asset management

Activities	Outcome
Municipal accounting	
1. Political decision early 90s	1. Introduction of double entry accounting
2. Automation of DEAS (double-entry account system)	2. Financial statements and ratio analysis
3. Parallel accounting	3. Asset and liability analysis
4. Complete switchover to DEAS	4. Asset accounting
5. Integration of online system	5. Use of innovative mortgage using stadium as a collateral
6. Inclusion in the WAN and LAN	6. Normative-/performance-based realistic budget for capital projects
Revenue administration and resource base	
1. GIS application	1. Complete listing of property tax account holders
2. Listing and objective assessment (unit area method)	2. Timely billing
3. Online collection through e-Seva Kendras	3. Quantum jump in nos and collections (Rs. 167 crores in 2004–2005 to Rs. 775 crores in 2014–2015)

Source AMC budget and discussions and IIPA study on urban governance reforms 2017

application of GIS in the municipal accounting and revenue administration in Ahmedabad Municipal Corporation (AMC).[9] It is important to observe that accounting reforms covering WAN and LAN to access statements and reports have enabled efficiency and transparency in the financial management and prompt transfer of money through RTGS, NEPC and ECS. This has built client satisfaction and confidence leading to reduction in bidding amount and timely completion of projects.[10]

AMC has also initiated community consultation through crowd sourcing using municipal website and WAN and LAN for (ii) placement of information on budget, assets, tax revenue. (iii) It has also improved AMCs resource potential covering the municipal rating and database on assets (Box 3).

Box 3
AMC Resource Potential

(i) Rating of municipal government by CRISL/ICRA to build investor confidence and repayment and borrowing capacities among select cities.

(ii) Expansion of tax base due to use of GIS. It has unlocked the tax potential and also developed a base for application of other land-based

---

[9]IIPA study on Urban Governance Initiatives in Ahmedabad and Bengaluru.
[10]AMC budget/Annual Report.

> tools and enforce value capture to recover the cost and raise funds for municipal infrastructure.
>
> (iii) Effective analysis of demand covering user-/slab-wise analysis of account holders of water supply and property tax. One can more effectively target the revenue base for optimum collection of municipal dues.
>
> (iv) Property tax innovations covered use of GIS for complete listing and verification and collections through online system in a decentralized manner.

6 E (Governance)-Service Centres (ESC)

E-service centres in Ahmedabad popularly known as Nagarik Seva Kendra are worth noting as another important example of crowd sourcing for municipal governance. 14.5 million visitors have used CSC services or seen the system in 2013–2014. ESCs have a 24 × 7 complaint registration system recording over 500,000 complaints in 2013–2014 out of which 80% are attended on time. Third-party assessment of ESC was done, and it has won awards from Indian Institute of Management, Ahmedabad (IIM-A), Government of India, and International City Managers Associations (ICMA).

These centres have yielded appreciable results. It may be seen from Table 5 that crowd sourcing under ESCs cover issuance of birth and death certificates, building plan approval, shops and establishment licences, registration of complaints, online tenders, hawker licence, restaurant licence, right to information, hall booking, TDR fees, vehicle tax and video conferencing. It is also noted that ESC 64 wards/55 civic centres are using a WAN network with 2500 computers at different locations and have contributed to a multiple increase (nearly 19 times) in revenue from Rs. 68 crores to Rs. 1236 crores during 2002–2015.[11] The increase in revenue confirms a quantum jump in revenue proceeds.

Similarly, crowd sourcing for property tax in Ahmedabad has facilitated widening of base and coverage along with collections. As may be seen from Chart 1, GIS applications has enabled complete siting of properties, and network of e-service centres has enabled online collections leading to substantial gain in the revenue yields.

---

[11]Municipal Budget of AMC and Annual Reports 2016 and 2017.

Table 5 E-service centre for municipal functions at AMC

Activities	Outcome
1. Integrated database management	1. City corporation website
2. Wide/local area network (WAN and LAN) 2500 computers for WAN	2. Wider e-coverage (AMC central office, 64 wards/55 civic centres)
3. E-service—(a) employee/office processing, (b) revenue, (c) operation and maintenance (O&M), (d) general data, (e) health, (f) town planning and (g) projects	3. 204 types of complaints for 24 departments
4. Complaint redressal—(a) in person, call centres, (b) online, (c) email and (d) through SMS	4. Nearly 500,000 complaints registered in 2013–2014
5. Inventory management	5. 80% redressal within time
6. Financial accounting system (FAS)/payroll	6. 24 × 7 registration access
7. Hospital registration	7. 1.8 million transaction in 2013–2014
8. College admissions	8. Third-party assessment
9. Video conferencing	9. Awards of IIM-A, ICMA excellence from National Award (2008–2009)
	10. Quantum increase in revenue (PT and others)
	11. 145 lakh visitors of CSC

Source AMC

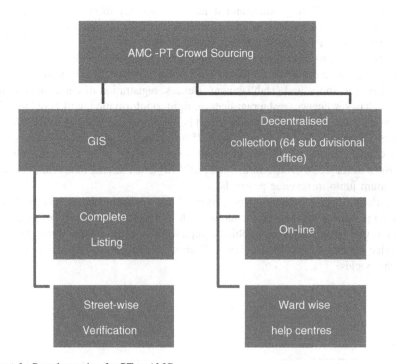

Chart 1 Crowd sourcing for PT at AMC

7 Management of Municipal Roads, Solid Waste and Assets

Application of crowd sourcing also includes other important O&M functions such as municipal roads, solid waste management and related assets which have undergone remarkable improvement among cities. E-governance/Geo-based smart delivery municipal roads in Bengaluru are one of the model approach initiated in the country.

Geo-based smart delivery of service applied by Bruhat Bengaluru Municipal Corporation (BBMP) includes (1) roads (2) land, space and property information, whereas innovative solid waste management covers dedicated links in the city website and use of Whatsapp groups to rationalize management of waste (Charts 2 and 3).

Geo-based application covering development of road inventory has facilitated development of consolidated inventory, optimal road asset planning and allocation, elimination of duplication of road works, real-time, consistent and descriptive information, business process optimization, work flows, status updation covering monitoring, quick execution, automated work flows, status updation for citizen and work provider, GPS-enabled mobile handlers and on-schedule data updates.

Geo-enabled land, space and property updates have covered property mapping, ward-wise grids, identification of properties not covered under municipal taxation, tax paid and defaulters, listing of municipal/assets, roads, lakes, parks and open plots/area having railway lines and wards/zones boundaries.

Solid waste management by BBMP is using crowd sourcing in a highly innovative manner. There is a city website which has seven sectors, 26 departments out of which exclusive link for waste management is given covering the system, network and grievance redressal procedures.[12] In addition, BBMP has facilitated Whatsapp groups for stakeholders, namely private vendors, community volunteers and municipal staff to interact for efficient delivery of waste management services. It has significantly helped to apply 3R (Reduce, Reuse and Recycle) waste and convert the waste as a meaningful resource (Chart 3).

Similarly, application of Geo-enabled land space and property information have enabled UID no's on properties, town planning, monitoring, revenue collection and monitoring of solid waste management. This has improved environment and enabled city to initiate innovative and participatory schemes such as Akrama Krama and My Bengaluru My Contribution.

Geo-based asset listing has enabled city to identify assets to apply suitable measures to improve upkeep and development of respective assets in a participatory manner. This is also benefitting city to have environment-friendly, pro-poor, cost-effective and inclusive development (Table 6).

---

[12]http://bbmp.gov.in/home.

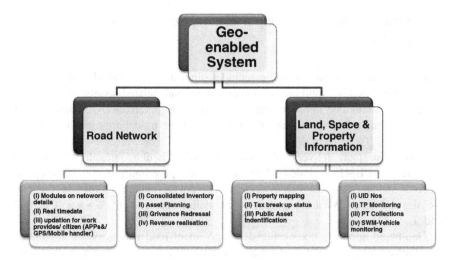

Chart 2 Geo-enabled smart delivery of roads, land, space and property information in BBMP

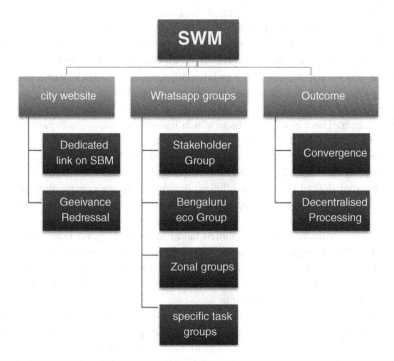

Chart 3 Innovation in solid waste management (SWM)

Table 6 Crowd sourcing for management of municipal roads, sold waste and assets

Activities	Outcome
Geo-enabled smart delivery of services	
(a) Roads	
1. Development of modules on road inventory, history, new road works, road cutting management, optical fibre cables (OFC) and laying management	1. Consolidated inventory
2. Business process optimization, elimination of duplication of road works	2. Identification of optimal road asset planning and allocation
3. Status updation for citizen and work provider	3. Employee productivity assessment
4. Apps on GPS-enabled mobile handlers and	4. Automated workflows
5. On-schedule data updates	5. Grievance redressal
	6. Improved revenue realization (expenditure efficiency)
	7. Real-time consistent and scripture information and, monitoring and quick execution
(b) Land space and property information	
1. Property mapping, ward-wise grids	1. UID no. on properties
2. Properties as per tax status (paid/unpaid and covered)	2. Property tax collection improvement
3. Earmarking public assets—road, lakes, parks, open plots	3. Town planning and monitoring
4. Identification of railway lines, ward, zonal boundaries, maps	4. Solid waste management (tracking vehicles through GPS-enabled services)
(c) Crowd sourcing for solid waste management	
1. Website specific links	1. Wider awareness
2. Whatsapp groups—separate groups for community, municipal officials, civil society and vendors	2. Wider participation of stakeholders— nearly 40 percent domestic waste is segregated at source
	3. Improvement in the waste management— different types of decentralised treatment for wet and dry waste is done at ward and household and community levels
(d) Assets management/listing exercise	
1. BBMP has undertaken a city-wide listing of municipal assets existing in the form of public spaces	1. BBMP identified 1688 public spaces covering 1146 gardens, 256 grounds and 286 other open spaces
2. State govt. issued a notification as per UDD 202 MNU 2012 dates 22 November 2012	2. Listing provided opportunity to prepare suitable plan
3. The notification indicates parks, play fields, open spaces for public use under Public Space Preservation and Regulation Act 1983	3. It was also used for innovative asset maintenance to raise participatory funding including Namma Bengaluru Namma Koduge as above

Data from IIPA study on Urban Governance Initiatives, 2017

8 Outcome

As emerged from the preceding analysis, the crowd sourcing in the three sets of municipal functions has significantly improved the governance and demonstrates application of different governance norms. Some specific examples for an illustration are as follows.

8.1 Decentralization

Accounting (zonal entries), budgeting, revenue demand assessment (complete listing of property tax account holders—property tax assesses, timely billing and collections and wider e-coverage—AMC central office, 64 wards/55 civic centres in Ahmedabad).

8.2 Accountability

Real-time data on O&M of roads in Bengaluru and financial statements and citizen centricity (204 types of complaints for 24 municipal departments). Nearly 500,000 complaints registered in 2013–2014 and 80% redressal within time in Ahmedabad.

8.3 Transparency

Asset accounting financial information, ratio analysis, financial statements, asset and liability analysis in Ahmedabad, Geo-enabled information on works in Bengaluru on roads, waste management and assets. Information is available to a large chunk of citizen and citizen are also accessible to city governments through crowd sourcing.

8.4 Efficiency

Financial transactions (electronically) and e-service centres for revenue collection/tax payments Quantum jump in no. of assessment and collections (Rs. 167 crores in 2004–2005 to Rs. 775 crores in 2014–2015) in Ahmedabad and employee productivity assessment/automated workflows and app-based services in Bengaluru.

8.5 Equity

e-service access to citizens across the Ahmedabad, citizen-centric asset management (parks, lakes, community centres) in Bengaluru.

8.6 Civic Engagement

Grievance redressal in Ahmedabad and stakeholder participation in the management of assets in Bengaluru such as health/fitness centre in Parks with support from private sector and local community (Namma Bengaluru Namma Koduge: My Bengaluru my contribution).

9 Follow-Up

As may be seen from above analyses, the crowd sourcing is emerging gradually as a tool to improve urban governance. Yet, the scope of expansion of crowd sourcing in terms of horizontal (across the cities) and vertical (within city and across the functions) growth is fairly wide. Rapid growth of access to mobile services, expansion of broadband services and municipal competition for reforms are major driving forces. Further, municipal initiatives have emerged as a result of tripartite consensus between centre, state and city governments to initiate access to crowd sourcing. It is also strongly felt that nodal ministries and departments in centre and states have to consolidate crowd sourcing and rationalize their respective states to more effectively use crowd sourcing as a tool to improve urban governance. Therefore, intergovernmental action agenda need to be developed for wider consensus and application.

Municipal models of crowd sourcing in the some of the key functions namely accounting and revenue administration, e-governance and management of roads, solid waste and assets have shown substantial improvement in the governance. It is, therefore, evident from above that ULBs need wider sensitization to improve their functions through crowd sourcing. Specific points for intergovernmental actions may include:

(i) Recognize further the role of crowd sourcing in the application of governance norms for more effective delivery of municipal functions. Accountability, decentralization, transparency, efficiency, equity and civic engagement can be improved significantly through real-time data and access to a range of stakeholders.

(ii) Centre and state governments should engage ULBs further to assess their progress, improve their formats and provide handholding to apply crowd sourcing at different levels, i.e. within city governments, across urban institutions and community/citizen interface.

(iii) States should develop a generic framework of data (basic idea, usability and creativity, technology design, and efficiency and quality) and its reporting as well as typology of mobile apps and city website to facilitate stakeholders. A broad guideline at national level in line with municipal accounting code (which is developed for financial management by centre) should also be developed. These need to be created in the form of manuals/checklist/ guidelines to systematically apply crowd sourcing covering website, links, apps, data entry/collection framework and citizen engagement process.

(iv) Carryout hackathon in a participatory manner covering state-level municipal administration department, city govt., participation of agencies and private sector (including students) to develop appropriate structure for reporting real-time data, its processing and access to consumers/beneficiaries of municipal services.

(v) All the city governments should create a city government website and establish link with internal (financial, O&M and investment) and external (community, vendors, NGOs) sources of information through social media, mobile computing, visualization using analytics and cloud.

(vi) Various municipal functions such as roads and transport need, sanitation and other services need to be linked with real-time data for accuracy and outreach in a most effective manner.

(vii) Financial management data (revenue, receipts/expenditure, assets/liabilities) need to be developed/updated online and accessible to stakeholders through Internet-based applications.

Effective Business Development for In-Market IT Innovations with Industry-Driven API Composition

Biplav Srivastava, Malolan Chetlur, Sachin Gupta, Mitesh Vasa
and Karthik Visweswariah

1 Introduction

Innovation refers to uniqueness in an entity's products, services, processes, or business models with which it serves its customers. It is well known in business world that IT-driven innovations can help a company differentiate its products and services with competition and drive business success through improvements in customer satisfaction, efficiency, productivity, quality, etc. [1, 2]. Some examples are manufacturing where adopting just-in-time (JIT) practices can remove inventory, logistics where an active RF-ID tag-based asset tracking can reduce delays or travel industry where social media can be used to engage customers to improve customer service and attract new market segments.

With millennial generation customers, continuous innovation is necessary to have them continue to stay with the existing services and products. In-market innovation is employed by businesses to both understand the adoption of their services and also to constantly improve the features based on the customer feedback and usage scenarios. This in-market innovation requires agile innovation practices and also has to respond to markets that are dynamic. Previous models of innovation involving longer cycles and catering to captive markets are long gone. Recent in-market innovation requires innovative frameworks to support agile innovation while firmly grounded on the market needs and value addition to existing services.

B. Srivastava (✉)
IBM Research, Yorktown Heights, NY, USA
e-mail: biplavs@us.ibm.com

M. Chetlur · S. Gupta
IBM Research, New Delhi, India

M. Vasa · K. Visweswariah
IBM Research, Bangalore, India

© Springer Nature Singapore Pte Ltd. 2018
U. M. Munshi and N. Verma (eds.), *Data Science Landscape*,
Studies in Big Data 38, https://doi.org/10.1007/978-981-10-7515-5_12

Despite the imperatives, businesses find it difficult to make go-ahead decisions on innovative solutions because, by definition, the innovations have not been field tested enough before to become off-the-shelf offerings and thus provide proven cost versus benefit business case. The decision becomes even more difficult if one is trying to purchase the innovation from an external entity like a start-up, university, research laboratory, or non-competing company whose ability to deliver is unknown. They want to know what scenarios can an innovation impact, what will be the business case seen from the point of view of new investments needed and benefit realized, what would be the assumption about other business processes and data to realize the promised value, what are other benefits beyond financial numbers (e.g., strategic, future ease) [3] and what are the sources of risk to guard against.

In the same setting, the providers of innovative solutions also face the problem of communicating business value while making their technology available for demonstration. Although they know the technology best, they do not have the knowledge about the proprietary customer data and business processes to estimate their technology's impact on customer's environment. As a result, they need to convince the potential client on the technology, use-case as well as business case to get the right price for the innovation and turn the client into a satisfied future reference. They also want to be cautious in parting with the details or actual solution without adequate compensation for their intellectual property in the innovations.

The problem we consider falls in the realm of business development for innovations which has received very little scientific attention. For regular IT-driven changes, it is commonplace to analyze a business using value maps and business operation network [4]. As an example of how to use it, in [5], the authors look at the banking industry and the factors which are transforming it globalization, innovation, and competition in the market and how the companies are responding.

To resolve the twin sides of the business development issues raised with innovation, in this paper, we propose an approach for quick prototyping of innovative solutions in the context of industry business processes and metrics by exposing technological capabilities as application programming interfaces (APIs) and assembling prototypes using API composition techniques. We call it accelerating client exchange (ACE) and discuss how it resolves the key issues in business development for IT-based innovation. Although there have been many a prior work on business processes modeling and service composition, they have focused on mainstream business operations and assume that the IT requirements are firm. In contrast, for our case of quick prototyping innovations and discovering business value in-market, to the best of our information, they have not been extended and applied for business development for innovations.

In the remainder of the paper, we first give a background of business development and then motivate our work with an example on how a technology can be used in different business contexts. Then, we derive a problem, provide the ACE framework as a solution, discuss a prototype, and demonstrate its usage. We conclude the paper with a discussion of related work, our limitations, and future directions.

2   Preliminaries

In this section, we first give the background on business development process for innovative technologies and then go through a case study of a technology innovation which can then be applied in multiple business scenarios.

2.1   Business Development Process

The business development process for an innovative technology can be broken into four phases: outreach, diligence, contracting, and in operation (see Fig. 1). Each of them consists of multiple steps.

A. Outreach:

1. Client identification—One forms a list of prospective accounts whom a vendor feels are potential targets. The challenge here is to nail down exact client names and divisions who can use the technology.
2. Client connection—Here, one develops contact with the client either through a known source or by a cold call.
3. Client meeting—One schedules meeting with the client with appropriate stakeholders and people from both sides. The challenge here is to know upfront who will this be most relevant to in the client.

B. Diligence: After outreach, one conducts due diligence with the client.

1. Discussion with client—Here, one undertakes open discussion on client businesses, issues, where they need help and what business metric improvement they seek. Technological innovation is usually not an end in itself but a tool to achieve a non-incremental business outcome.

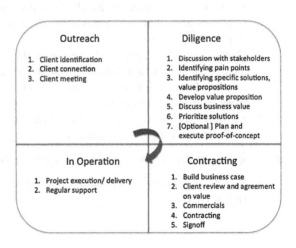

Fig. 1 Business development process for innovations

2. Identify pain points—Post discussions, a few pointed areas of collaboration are identified to tackle pressing problems.

3. Identify specific solutions, value proposition—Usually, there are innovation choices for the identified problems. Working with client explores their feasibility by checking assumptions and verifying with any sample data available.

4. Develop value proposition—Here, one goes beyond the system (input/output) view of the technology to understand the context (tools, procedures, people, regulations) in which it will be used at the client end and list out the impact. The impact constitutes the cumulative value from the technology for the client and needs to be validated with the client.

5. Discuss business value—Here, one collaborates with the client to develop a business case with anticipated potential benefits to the client and the requisite investment needed.

6. Narrow down/prioritize solutions—Now, the identified solutions are prioritized by mapping them with key dependencies and consulting the stakeholders.

7. (Optional) Proof-of-Concept (PoC) prototype—If requested by client, one also needs to plan and executes a PoC prototype where a scaled-down demonstration of the technology is done for the customer to test pre-identified hypotheses.

C. Contracting: Now one negotiates a contract to deliver the innovation to the client.

1. Build business case, cost–benefit analysis—Here, one articulates what the client will benefit from the technology in their business and also what investment is needed. Special attention is also paid to anticipated risks and ways to overcome them.

2. Review with client to agree in-principle on value—Now, one convinces the client to agree with value jointly developed. The client may want additional factors to be incorporated that they had not realized at the outset.

3. Commercials—Here, all parties agree on indicative pricing with timelines, responsibilities, and milestones.

4. Contracting—Now, approvals are obtained by the technology vendor from their sub-units who look into the legal, financial, and strategic aspects of the contract. Once all internal approvals are obtained, the contract can be put on paper in front of the client.

5. Sign-off—the client would also do their due diligence before signing the agreement to adopt the innovation.

D. In operation: Putting the innovation in client's environment.

1. Project Execution—Here, one delivers the technological innovation as per signed contract.

2. Support—One now does regular project management and support with governance to ensure the customer realizes the promised benefits.

2.2 Motivating Scenario

We present a common scenario to motivate the problem. It consists of a technology innovation which may be applicable to multiple industries.

Innovation Example—Linking Enterprise and Social Media Data. Organizations of all types maintain data about their operational activities in enterprise databases. An example of such data is information about an organization's employees, customers, partners, and shareholders. Now some of these entities are also active on social media channels like Facebook, Twitter, LinkedIn, and YouTube expressing both personal and business opinions. However, they may not be using the same identity to interact with an organization as they do on social media.

In this context, one can use technological innovations to link a person whose profile is known and stored in enterprise data against the online profiles available on social media. Such a link can help a business gain deeper insight about its existing customers leading to better suggestions for its product and services. We will call the capabilities social linking, and it consists of two cases:

- Given two datasets related to people, an enterprise dataset with primary keys and a social dataset with identifiers (keys), find a (partial) mapping between the primary keys in the two datasets.
- Given an enterprise dataset and a social dataset related to people, and known mappings between their primary keys, find significant topics [6] related to people spanning structured and unstructured data in the two datasets.

This is an area of much ongoing research and innovation, and a further variation involves linking open data as well [7]. We now show some examples of using this innovation in mainstream and new industries. See [8] for more examples and details.

Application Example 1

Purchases by a bank's account holders. Finance is a well-entrenched industry. Here, banks maintain extensive information about their account holders as part of government mandated regulations (e.g., demographics) and normal operations (e.g., credit and debit to accounts). They are always eager to get a higher share of their customer's wallet with their portfolio of financial offerings.

In this context, a bank can use the social linking innovation on its account holders and data about their social activity wherein the bank's offerings could be relevant. The bank could even customize their products if they see an unmet demand from this analysis. Any leads which the bank gets can be seen as improving its marketing and promotional operations.

Application Example 2
Increasing online booking for an airline. Like finance, transportation is a well-entrenched industry. Here, airlines book customers on their flights via their online Websites, booking cum sales offices and also through third-party travel agents as well as partner airlines. Among these options, booking customers via their Websites are the most desirable because it is not only the cheapest to operate (per ticket sold) but also the one giving the highest revenue (per ticket sold since no portion of the fare has to be shared). Hence, airlines would like to shift more of their bookings to their online platforms. In the travel context, the airline can use the social linking innovation on its previous fliers, whose data it already has in enterprise databases, and data about their social activity, to highlight flights, travel packages and offers that will match their expressed or predicted needs. Indeed, this will meet the airline's objective of getting more business from their established clients while also improving the customer's perceived satisfaction from the airline due to their customized overtures.

Application Example 3
Tracking people of interest for safety and security. Public safety and national security are high priorities for government's worldwide. Here, they build profile of people that are of interest and want to continuously update them. Private businesses are also sometimes interested in tracking people to avoid theft and disruptions to their operations.

In safety context, agencies can use social linking innovations on its citizens and visitors to build a holistic behavioral profile and prevent untoward incidents in a timely manner. The biggest practical challenge (and opportunity) for innovations is adhering to data privacy laws across multiple agencies (public, private) and countries.

Example 4
Increasing student enrollment in online courses in education. Innovations in online education are aiming to transform the way we learn with Massive Online Open Courseware (MOOCs) [9] in higher education and lifelong learning segments. Here, online education providers have to design curriculum that is relevant to industry needs, consists of appropriate learning pathways leading to appropriate skilling, and leverages customizations based on the existing courses (content, experts, ecosystem of learners) [10].

Existing information on existing courses and learner population and experts can be obtained from the enterprise data maintained by the online education providers. The information about the job opportunities is available in open (from job advertisements and blog posts of companies) and from the social data posted by various companies. The learner profile and their biography can be obtained from both internal information about the learner and their social profile from LinkedIn, Facebook, and other social communities. In order to improve course enrollment, the courses offered must demonstrate measurable outcome in terms of new opportunities or career progress of past learners to current learners. These could be determined from the changes in their profile before and after the course completion from the social data and references to positive sentiments on the course offerings (if any). Therefore, to enable in-market innovation of new course offerings relevant to changing workforce requirements, there has to be a constant feedback of social data about the skill demand and change of profile of learners toward customized offering.

Summary in above, we presented just one example of a technology innovation (on linking enterprise and social data) which could be applicable in finance, travel, government, and education sectors. A technology organization would be looking to bring 10–100s of such innovations to market quickly, and we want to enable their business development.

3   Problem

The problem we seek to tackle is that given a technology

1. demonstrate that it works in the advertised setting(s)
2. show that it can work in the client setting
3. show that it can improve the business metrics of client interest
4. without revealing the internal working of the technology

The above scope will help the business development process in the following phases and steps:

- In diligence phase, facilitate ease in #3 (identifying solutions), #4 (developing value proposition), and #7 (conducting proof-of-concept prototype)
- In contracting phase, facilitate ease in #1 and #2 (building business case and reviewing) as well as justifying #3 (commercials)
- In operation phase, enable better project execution

For the rest of the paper, our running example for the problem will be social linking innovation for the finance domain. As [8] points out, the business case for innovations in this space is still not clear, and hence, co-creation is the best way to engage customers.

4   Solution

The outline of our solution is the following:

1. Leverage industry-accepted ontology of business processes and metrics as the lingua franca for business development communication.
2. Enable capabilities and data in scope, both innovative and preexisting, as APIs.
3. View market-wide use-cases clients are talking about as business processes implemented by IT applications that are themselves assembled by composing APIs.

4. Execute interactively and live while engaging the client using the API platform.
5. Discuss business metrics for evaluating impact.
6. Take go-ahead decisions proactively and transparently to improve client satisfaction.

In Step 1, we start by using an industry-wide ontology. We will use American Productivity and Quality Center (APQC) process classification framework (PCF [11]) as the industry-accepted ontology of business processes and metrics. APQC's PCF, APQC for short, is a standard for terminology on process definitions and measures for benchmarking. The development of APQC taxonomy started in 1992. More than 80 organizations have helped it evolve, and it is now being used by thousands of organizations worldwide. APQC is organized into five hierarchical levels representing business functions at different levels of granularity categories, process groups, processes, activities, and steps. To convey the level of granularity, roughly put, a level five tasks can be done by an individual, a level 4 activity is done by one or a group of individuals, and a level 3 process is accomplished by an organizations group. Companies organize functions at level 1 and level 2 in many ways depending on size, geography, and other considerations. At each node, there is a number in the bracket referring to a unique identifier assigned in BPH. APQC offers one, industry-neutral, hierarchy (also referred as cross-industry) plus many industry-specific frameworks to describe processes (or process nuances) specific to the industry. Note that a company might be identified by more than one industry BPH, depending on their profile.

As illustration in Fig. 2, APQC's business processes from the finance industry are shown (version 6.1.1). Our example use-case falls under 3.0 category of business processes dealing with selling of products and services. This is further scoped down to 3.4.6 for promotional activities. The metrics related to the business process are shown on lower right based on their match using the Process IDs (10,152 and 10,167). Thus, any company wanting to use linking innovation should be looking at improving one or more of these business metrics.

In Step 2, the access to data and capabilities is viewed as application programming interfaces (APIs). APIs are popular building blocks for open-standards-based IT today. Here, service providers publish specification of their IT capabilities wrapped as services onto registries. The services can be discovered by potential consumers later and then invoked on the providers, all using standardized interfaces. The idea has been popular as service-oriented architecture (SOA) which could run with proprietary technologies, as Web services with their focus on XML format for data serialization, and today using representational state transfer (REST) and JSON data format. We use a common term, Web APIs, or APIs for short, to refer to them.

In Step 3, the use-cases of client interest are viewed as applications that can be assembled from APIs. The API platform, thus, is used to encapsulate sample data, compose using unique innovations and preexisting capabilities, to trigger necessary decisions (visualizations, actions). There are many choices for composition techniques to use [12, 13], and the most popular is to do composition on the glass.

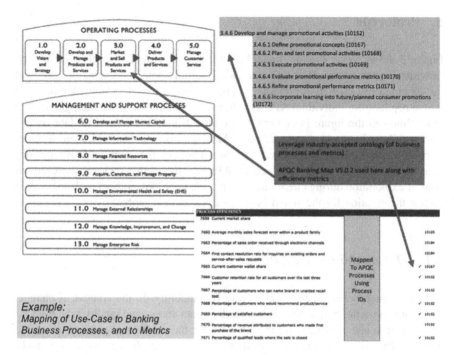

Fig. 2 A finance use-case with social and enterprise data

In Step 4, the innovation provider and client explore the different situations in which the innovation can be used. Some of them may not have been discussed earlier, but now seem feasible and the client would want them to be considered. For example, in our running example of finance, the bank may want to offer not only the finance products to its existing customers based on their social expressions but also (travel) insurance products that it had little insights about earlier for its customers.

In Step 5, the client and innovation provider discuss the metrics that should be the focus for determining business success. Here, there are many models which finance stakeholders (e.g., Chief Financial Officers) prefer like return on investment (ROI) or payback period [14] while others that technical stakeholders prefer (e.g., Chief Information Officer) like improvement in productivity, decreased costs, and added flexibility with some overlap. Even when the two stakeholders may agree on a metric like ROI, they may put different emphasis [3].

Finally, since the client can see the innovation in action, albeit in the small, and understand their potential as well as assumptions, they can take go-ahead decisions effectively. The innovation provider will also be confident that the client has co-validated the solution to meet their business goals, thereby increasing the chance of their initiative's success. APQC provides industry benchmark values for the most common metrics, and this can help set unambiguous objectives.

In the next section, we show how these steps come together in a prototype we have developed to validate these ideas.

5   Prototype

Figure 3 shows a prototype for our solution from the point of view of an innovation provider who wants to pitch innovations to multiple potential clients built around the example use-case. It follows the composition on the glass composition style where results from individual APIs are shown on a common portal (i.e., glass). The left column in the figure is dedicated to output from APIs related to data access. The middle column shows output from APIs related to technological innovation. The right column depicts the results in the context of the overall APQC industry map and other innovations listed on an API registry so that they can be used for what-if exploration during client interactions. We now look at them in detail as they realize the solution for the stated business development problem.

5.1   IT Capabilities as APIs

Data is a crucial element for running any IT capability. In the context of change, i.e., innovation, there may be issues regarding data access and quality which varies from client to client. Hence, in the prototype, we have APIs that connect to sample data on which the innovations can be demonstrated to run. In the sample data itself, the schema is important while the content is illustrative with personally identifiable

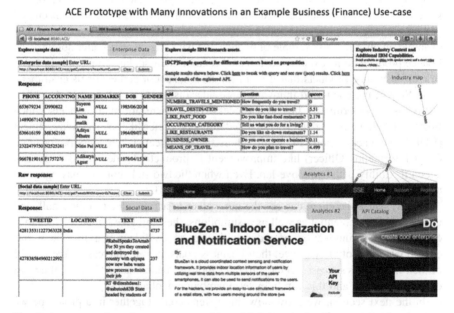

Fig. 3 A prototype for a finance use-case with social and enterprise data

information anonymized. The data-related APIs further help to transparently highlight dependencies which need to be captured in subsequent business value discussion. In our prototype, we provide sample enterprise data in the form of demographic information for a set of hypothetical people who may be a client's customers and social data in the form of Twitter updates by another set of hypothetical people, with some overlap.

Another set of APIs relates to demonstrating the IT innovations in scope for a client. Each innovation is available as an API which exposes multiple endpoints. The prototype calls specific endpoints in the innovation with default sample data which can be overridden as the client interacts with it in a live session. In the screenshot, two other innovations are shown available for exploration [15] in addition to the example one on social linking. The advantage of using APIs to demonstrate innovations is that one does not have to install the technology at the customer premise or reveal details of the internal workings, both of which help in preserving the innovations' intellectual property.

Finally, a low-level detail worth highlighting is that the results from the APIs are obtained in JSON or in XML format or different visualization formats. The prototype provides converters to turn them into human readable renderings.

5.2 Business Value Calculation from Innovation Investment

As noted earlier, there are many investment evaluation methods preferred by financial and technical (IT) stakeholders in businesses [3]. However, despite the divergence, return on investment (ROI) is the most common among them [14]. It is defined as:

$$\text{ROI} = \frac{(\text{Benefit} - \text{Cost})}{\text{Cost}} * 100 \tag{1}$$

We provide a template calculation for ROI where components for benefit include metrics mapped to the 3.4.6 business process (and its sub-activities), their targeted improvement (a percentage which the customer can override), and the financial interpretation of that improvement (a formula to arrive at $ impact) for the client. Similarly, the components for cost include the cost of making the data available for innovation, the change to client's existing IT systems to store new results, and the cost to procure the innovation.

The client can look at the ROI components, tweak the default values, add more components that they deem necessary, and get a quantitative indication of innovation's impact. All assumptions and default values in the calculation are highlighted and referenced to industry baselines from APQC.

5.3   APQC Industry Map as Interactive Map

Although the focus of the ACE demonstration is around a particular use-case (i.e., finance), the client may want to explore other related business processes, metrics, or other capabilities that the innovation provider may have beyond our example innovative solution (i.e., linking enterprise and social data). Therefore, we provide a novel interactive visualization in the form of a radial map (see Fig. 4) in the prototype. The client and the innovation provider can interact using the APQC PCF finance industry ontology to explore different what-if cases. As the user scans the processes, the full names emerge as also does the matching business metrics. Here, process identifiers from APQC are used for correlation. Further, industry-specific benchmarking data from APQC can be used to establish what the client objective may be with respect to their competitors.

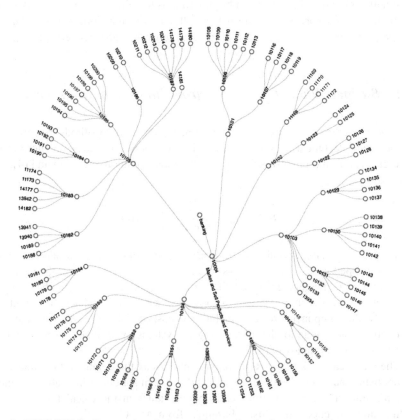

Fig. 4 APQC PCF for finance viewed along with mapping business metric(s)

5.4   Industry-Customized Functional API Search in Registry

Like the use-case, the client may also want to know what other capabilities the innovation provider has. Hence, we allow the ACE prototype to access an API registry where in addition to basic information about APIs that are captured in a typical registry [16] (i.e., name, description, endpoints), the APIs are also tagged with industry-specific or cross-industry business process and metrics that they can influence. Thus, in the context of the business process and metrics the client is interested in, we can search the registry for other related innovations. This not only enables a thorough discussion with the client but also provides additional cross-selling opportunities that business development is always interested in.

A demonstration of API search is shown in Fig. 3. Here, when the client wanted to see capabilities beyond social linking as applicable for a financial firm for marketing, innovations related to contextual questioning of customers and tracking in closed spaces [15] are discovered and shown.

5.5   Early Experience

Recall that the impact of our approach would be in diligence phase in the steps of identifying solutions, developing value proposition, and conducting proof-of-concept prototype; in contracting phase in facilitating ease in building business case, reviewing with client and justifying commercials, and in better operational execution. Prior to using the new approach, the critical steps in business development process with clients would involve showing video demonstrations of innovative technologies, questioning the client on their pain points, understanding their business process and data to estimate the scope of changes, setting up a proof-of-concept demonstration to justify a business case, and then concluding a contract. The efforts would span proving the new technology, understanding the client's industry and their unique situation, and establishing a business case for innovations of interest. It was typical to spend months completing all the whole process.

We have started applying our solution as implemented in the ACE prototype for finance industry with select clients offering a range of innovations with promising results. As the innovations are demonstrated to run live with sample data, there is minimal effort spent in proving the technology. Further, a standard industry ontology is used to discuss business processes, metrics, and baselines, understanding client's unique situation is simpler. Consequently, most of the time during the business development process is spent in building the business case—a drastic reduction in end-to-end business development time. We have been able to respond to potential clients in days where it was taking months earlier.[1]

---

[1]Note that due to business reasons, exact case studies cannot be shared.

Finally, it is easy to pursue an innovation in multiple industries by switching to corresponding industry maps that contain relevant business processes and metrics. ACE's underlying framework for composing APIs, allowing browsing of the industry, searching the registry of APIs with industry tags and business value calculation remains unchanged. This will come handy in exploring clients for the example innovation in the transportation industry and new industry like education in future.

6 Discussion and Related Work

We now discuss how our work connects to prior work and can be improved further.

6.1 Related Work

The prior work related to our problem can be found in management literature around business strategy and processes and (CS/IT) services literature around Web services and APIs.

It is commonplace to analyze a business using value maps and business operation network. As example, in [5], the authors look at the banking industry and the factors which are transforming it—globalization, innovation, and competition in the market. They then discuss an integrated approach for business network redesign for the financial industry that covers the layers of strategy, business process, and operational (IT) systems. However, there is interest to know how innovation impacts a business. In [1], the authors look at why some companies do exceedingly well compared to their competition and articulate the reason to be how they question the assumptions of the industry and provide new products or services that disrupt the traditional value chain. To do so, the successful companies embrace technical and business innovation which leads to a quantum jump in revenues and profits compared to the competition which is still in incremental improvement mode. In [2], the authors discuss the experience of a large research-driven organization, Xerox, on how they fared in realizing value and success from innovative technologies. Specifically, they highlight that apart from the technical ingenuity, new business model may also have to be explored which is in line with client's perception of value.

In services literature, tieing business processes and objects with their service-oriented architecture (SOA)-based IT implementation is well understood. Packaged integration platforms like IBMs composite business services [17] or SAP have pre-configured business processes that are implemented and exposed as services. Using them, service composition [13] can help accelerate time to value for a business and service customization can help handle changes [18]. However, there are choices which must be considered judiciously. In [12], the authors proposed that common service composition techniques fall into six patterns with distinct characteristics of their integration intermediary: (1) composition on the glass, or Type G, where services are connected by presenting their user interfaces on a screen and having a human operator serve as an intermediary to route connection messages; (2) composition by widgets, or Type W, where services are connected programmatically by interactions between their on screen representations, such as forms, maps, and buttons triggered by humans; (3) composition by direct invocation, or Type D, where services are connected by developing a custom application that calls the services and handles the data and control flow explicitly; (4) composition by information sharing, or Type I, where services are connected by a shared access to persistent storage space which holds service invocation information; (5) composition by orchestration, or Type O, where services are connected by a workflow specification which defines both communications among services as well the behavior of the composed process; (6) composition by events, or Type E, where services are connected by publishing and receiving event messages to predefined topics on a pub/sub server, which serves as an intermediary to asynchronously deliver connection messages. They compare the patterns around ability to scale with transactions, ease in monitoring, robustness against exceptions, evolution to changes, and data interoperability, and find that no pattern is better than others in all business and IT situations. We chose composition on a glass in our prototype.

6.2   Limitations and Future Work

As previous sections noted, the ACE approach has shown that business development duration can be reduced by having API technologies and reusing industry knowledge. However, the current prototype is limited to finance industry and needs to be expanded. Further, there are many financial and technical models to evaluate IT change [3]. We focused on ROI, and one can incorporate others. The current composition approach in the ACE prototype is Type G (glass), and we can incorporate further automation by moving to Type W (widget). Additionally, one needs to gain more experience with a wider client base as well as have the approach tried by other innovation providers.

For the ACE approach to work, the IT innovation needs to expose an API endpoint. However, many innovations do not do this at the outset and converting them to support API interfaces becomes time-consuming. But evidence shows that doing so makes the innovation available to multiple usage contexts which were cumbersome to pursue earlier. As example, in ACE prototype, the social linking innovation can be pursued in multiple industries easily by switching to appropriate business maps—a capability we plan to incorporate in future.

7 Conclusion

Continuous in-market innovation facilitated by IT has become an imperative for business across industries. Motivated by the need to accelerate such innovations while balancing the concerns of both clients and innovation providers, in this paper, we proposed a framework, called ACE, to develop and incorporate innovative solutions in the context of industry business processes and metrics using (Web) API composition. We have implemented the ACE approach into a prototype, and early experience shows that it is able to expedite business development by either reducing the turn around time to contract or in identifying the mismatch between the features and the targeted market segment. Overall, the proposed framework has helped in communicating value to clients toward planned or proposed in-market innovation features.

Acknowledgements We will like to thank Rakesh Pimplikar and Srikanth Tamilselvam for their help in implementing the ACE prototype.

References

1. Kim, W. C., & Mauborgne, R. (1997). Value innovation: The strategic logic of high growth. *Harvard Business Review*, 103–112 (January–February issue).
2. Chesbrough, H., & Rosenbloom, R. S. (2002). The role of the business model in capturing value from innovation: Evidence from xerox corporation's technology spin-off companies. *Industrial and Corporate Change, 11*(3), 529–555.
3. Silvius, A. J. G. (2002). Does ROI matter? Insights into the true business value of it. *The Electronic Journal Information Systems Evaluation, 9*(2), 93–104.
4. Hines, P., & Rich, N. (1997). The seven value stream mapping tools. *International Journal of Operations & Production Management, 17*(1), 46–64.
5. Reitbauer, S., Kohlmann, F., Eckert, C., Mansfeldt, K., & Alt, R. (2008). Redesigning business networks reference process, network and service map. In *SAC*.
6. Granitzer, M., Kroll, M., Seifert, C., Rath, A., Weber, N., Dietzel, O., & Lindstaedt, S. (2008). Analysis of machine learning techniques for context extraction. In *Third International Conference on Digital Information Management, 2008. ICDIM 2008*, November 2008, pp. 233–240.
7. Omitola, T., Davies, J., Duke, A., Glaser, H., & Shadbolt, N. (2014). Linking social, open, and enterprise data. In *Proceedings of the 4th International Conference on Web Intelligence, Mining and Semantics (WIMS14). WIMS '14*, New York, USA, ACM, pp. 41:1–41:8.
8. Fan, W., & Gordon, M. D. (2014). The power of social media analytics. *Communications of the ACM, 57*(6).
9. Liyanagunawardena, T., Adams, A., & Williams, S. (2013). MOOCS: A systematic study of the published literature 2008–2012. *The International Review of Research in Open and Distance Learning, 14*(3).
10. OED: Open education alliance. (2014). https://www.udacity.com/open-ed.
11. APQC: APQC process classification framework (PCF). (2014). Available at http://www.apqc.org/.
12. Chang, Y. C., Mazzoleni, P., Mihaila, G. A., & Cohn, D. (2008). Solving the service composition puzzle. In *Proceedings of SCC*.

13. Agarwal, V., Chafle, G., Mittal, S., & Srivastava, B. (2008). Understanding approaches for web service composition and execution. In *Proceedings of the 1st Bangalore Annual Compute Conference. COMPUTE '08*, New York, USA, ACM, pp. 1:1–1:8.
14. Farris, P. W., Bendle, N. T., Pfeifer, P. E., & Reibstein, D. J. (2010). *Marketing metrics: The definitive guide to measuring marketing performance*. Upper Saddle River, New Jersey: Pearson Education, Inc. ISBN 0137058292.
15. Agarwal, V., Banerjee, N., Chakraborty, D., & Mittal, S. (2013). Usense—a smartphone middleware for community sensing. In *2013 IEEE 14th International Conference on Mobile Data Management (MDM)* (Vol. 1, pp. 56–65). IEEE.
16. Web, P. (2014). PW API registry. Available at: http://www.programmableweb.com/apis/directory.
17. IBM-Global-Services: Accelerating business flexibility, while reducing costs, with composite business services. (2007). http://www-935.ibm.com/services/us/index.wss/offering/gbs/a1027243.
18. Mazzoleni, P., & Srivastava, B. (2008). Business driven SOA customization. In *Proceedings of ICSOC*.

The Data that Get Forgotten

Elizabeth Griffin

1 Preamble

The word *Landscape* in the title of this book is a fortunate choice as it is particularly descriptive in the context of scientific data. If we turn ourselves around, wherever we are—in a city, in the country, among hills or on the sea, the landscape presents a continuous panorama until our view comes round again to where it started. It is an unbroken theme of connectedness, of relationships (tree to field, shore to ocean, river to topography…). Evolution, interpreted here as a development that interrupts those harmonies, is noticed when a building such as a tower or a spire stands out in front of the background landscape and obliterates the full view unless we move our sight-line. Such evolution can certainly be considered an integral element of a changing landscape, but it was not, or is not, its driver or builder. It has become imposed upon, but does not interact with, what is present in the background.

2 The Scientific Landscape

Ever since homosapiens became alert, observations have been made, facts and curiosities noticed and noted, and records kept in some form for future reference. Even if only featured as hearsay tales or non-numeric signs, 'observations' record what was detected by one or more of the senses—dominantly the visual ones but by no means exclusively so. Between them, those records create a panoramic landscape of information about the scientific world, and an undisputed witness to myriad changes within it, whether rapid or creeping, abrupt or subtle. Trying to understand the implications of an observation would invariably come to mean

E. Griffin (✉)
Dominion Astrophysical Observatory, Victoria, British Columbia, Canada
e-mail: Elizabeth.Griffin@nrc-cnrc.gc.ca

© Springer Nature Singapore Pte Ltd. 2018
U. M. Munshi and N. Verma (eds.), *Data Science Landscape*,
Studies in Big Data 38, https://doi.org/10.1007/978-981-10-7515-5_13

discussing and sharing relevant information, thereby seeding the culture of scientific research: an environment of sharing, comparing, discussing and co-building.

Scientists have always embraced that ethic of sharing, by adopting and building on what had already been learned, utilizing laws already proven and following conventions already woven into the intellectual environment. The foundations of pure research, sometimes called 'blue skies research', benefited very substantially from individuals who were able both financially and intellectually to run their own laboratories and to follow their own natural inquisitiveness, even the point of taking serious personal risks (the Curies clutched radioactive materials, Franklin nearly electrocuted himself with lightning, vaccines and medicines were tested on whoever...), but most were sole operators; those 'gentleman scientists' placed considerable importance upon communicating their results to the membership of a Learned Society, and were often themselves the founders of such Societies.

3 Building Libraries of Data

Those Learned Societies were both resource and meeting point. With their founders firmly behind them, many created libraries that assigned space for maps, charts and atlases, and several have remained prime or national repositories for such artefacts. Some created embryonic archives of collections, samples, and tools of the trade, though objects or records that were not easily or safely transportable were better stored closer to the scene of collection or analysis. Scientific data thus became part of the laboratory landscape, readily available in-house and able to be inspected or borrowed by visitors. That aspect of scientific research has not changed. Although recent observations are very largely electronic, and while libraries of paper products have been downsized and archives of charts and the like superseded by huge volumes of formatted files requiring powerful computer access, the need to inspect, consult, revisit and borrow data taken in the past has not fallen out of fashion. *Those observations and records remain an integral part of the scientific data landscape.*

Nevertheless, the question *Why?* is raised by surprisingly many scientists. Why trouble with 'old stuff' when we have so much more, and more superior, modern stuff? Science has tasted Progress, and the soulmate of Progress is Technology. Technology has a different landscape; it absorbs inspiration through evolution and notices only the arrow of time that points forwards. We do things faster, in more detail and in increased depth and precision compared to our predecessors in science. We 'observe' in a multitude of different modes, by ultrasound, X-ray and CAT scan, simultaneously from different locations, from habitable ground-levels, from cities, from distant areas devoid of civilization, from mountaintops, from the sea-floor and from space. Indeed, why struggle to re-read those old observations, given their intrinsic limitations—why even keep them since it costs precious resources to do so? Is the preservation game really worth the candle? If keeping is for keeping's sake, then surely it is only necessary to keep a few samples to illustrate bygone technologies. Why plan to keep them *all*?

In fact the answer is indisputable and lies in the date-stamps which those older records bear. No recent observation, be it ever so sophisticated, can reveal what the conditions in the sky, the atmosphere, the stars, the Sun, the sea, or any biological system or physical feature, were like several decades ago. Extrapolating today's data backwards by modelling the present is no substitute; it only reflects an extension of present trends and cannot bear a reliable approximation to actuality except by chance, and chance is never a substitute for reality. In just a very few cases, such as (for example) applying new methods to determine a chemical or physical constant more precisely, a lack of historic data may not matter because the constant concerned is just that—unvarying by definition, so new data supersede the older ones; there may be little virtue in keeping inferior versions except for historic interest, and small samples will suffice for that. However, the vast majority of situations concern either living objects such as populations of biological systems or specimens or migratory patterns, or evolving inanimate ones such as glaciers or volcanoes, stars that slowly consume and exhaust their internal stores of energy, varying concentrations of stratospheric ozone, seasonal cloud cover, river levels, the brightness of variable (and even of supposedly non-variable) stars—the changes which can be revealed by comparisons between past and present images can be not only stunning but may be of the utmost importance for divining the causes of the changes. Had the data never been kept and preserved, those various pieces of essential knowledge would never have been gained. The *only* route to that knowledge is therefore by direct appeal to historical ('heritage') observations and records.

4 When Heritage Data Prove Pivotal

Examples of new science that has been achieved in a variety of different disciplines through accessing old data can be found in the scholarly literature or popular journals. They fall roughly into two categories: (1) the results of painstaking and persistent investigations to recover and transform digitally data which were known to exist but could not be accessed (and sometimes not even located with any certainty), and (2) the stroke of luck, or serendipity, when the right person happened to be in the right place at the right time and chanced to hear mention of, or catch sight of, caches whose whereabouts a colleague or acquaintance had long sought. Quite often, though, the rescue is a hybrid of both kinds. As a collection, the data-rescue histories constitute a powerful resource for the free use of those who champion the preservation of heritage observations. Some represent the culmination of months or years of patient searching, assessing and re-analysis, often involving the resurrection or recreation of obsolete technology such as tape-readers or densitometers in order to handle the heritage data correctly.

Within each scientific discipline and sub-discipline, one encounters a great many examples in which quite intricate and detailed long-term variability has been recognized, decoded, and new science has emerged, be it the migration of species of flora and fauna under the external stimuli of changing environmental conditions, silting up or clearing of rivers through natural processes, changing tidal currents and the erosion of coastlines, or the way some apparently stable and unchanging stars show slow but persistent variations in velocity through the gravitational effects of a hitherto undetected companion star. A dynamic bibliography of research that has involved the rescue or incorporation of heritage data into modern research is being compiled by the *Data Rescue Interest Group* of the Research Data Alliance (rd-alliance.org/groups/data-rescue.html); it will be placed in the public domain, where additions will be gratefully accepted.

Since 'change' is the name of the game, one very significant agent of change in our present natural world is anthropogenic interference (however altruistic the intent). Occasionally, we can discern the hindrance to progress in research brought about by bias (see 4.3.2), particularly when the research in question is ahead of its time; other examples illustrate the value of being able to share data across traditionally impenetrable barriers between sciences, thereby strongly advocating a push towards the principles of Open Data and for working towards standardizing data formats. A few examples relate happy stories of pure serendipity (category 2 above). In addition, data archives, especially the oldest ones, tend to be eclectic: one is not quite sure what to expect, and the subjects of the data themselves suggest the research, rather than the other way about.

4.1   Systematic Rescue and Modernization of Heritage Data

4.1.1   Biodiversity Data

In the biological sciences, the objects themselves (the samples that are measured or recorded for study) can disappear or become extinct as well as the associated measurements and records, so preservation of information has two prongs: (a) while we cannot halt the progress of evolution, we can capture and save images of vanishing species if not already too late, and (b) we can ensure that all kinds of measurements made in the field or in the museum are properly preserved. In the biodiversity sciences, a large number of the activities of data recording are carried out by small groups or individuals (e.g. for a student thesis) rather than by routine monitoring by institutions. So often those theses, and the data written there, remain unpublished. Without a well-advertised and effective scheme to encourage and possibly assist the deposit and suitable homogenization of such data into public archives, the rates of attrition of those very valuable biodiversity data will continue and probably accelerate.

A team at the German Botanic Garden and Botanical Museum (Berlin-Dahlem) created 'reBiND' (rebind.bgbm.org), a three-year project funded by the German Research Foundation to develop cost-effective work-flows for transforming threatened biodiversity datasets into a well-curated, sustainable archiving system that feeds data automatically into appropriate trans-national information infrastructures such as GBIF (gbif.org) or BioCASE (biocase.org). As described by Güntsch et al. [1] and references therein, a wide range of isolated biodiversity data has already been rescued using reBiND and converted into new science; the topics which have benefited range from historical herbarium sheets from Hamburg Herbarium (Germany) collected during expeditions more than 100 years ago, to the vegetation on Swiss peaks in 1906 which are contrasted with the corresponding conditions 100 years later [2, 3] via specifics about the occurrences and extinction risks of Grand Beatles and Grass Snakes. Some rescued records reported by Güntsch et al. include morphological data on shearwater birds on Robinson Crusoe Island, Argentina [4], that were hidden in unpublished reports; others show how changes in the vegetation on sand dunes on a North German island has depended on grazing by wild ungulates [5].

The scales of some of those enterprises may appear to be small, but every element is critical to a specific bio-system, and every specific system is an indispensable link in the chain of systems that maintain the equilibrium of planet Earth. The patterns of change revealed by bio-systems are fascinating sources of information: the plants and animals sense what is becoming less (or more) favourable and migrate accordingly, and it is up to the researcher to discern the cause(s). ReBiND also created a short video cartoon, which made its message easily accessible by all ages and stages. *Moral: the more you tell people, the more people you can involve!*

4.1.2 A Central Archive for 45 Years of Landsat Data

The *Landsat* satellites, which have been imaging planet Earth from space for the past 45 years under the auspices of the US Geological Survey, provide unique and rich input for studies of (for instance) impacts of agricultural and forestry policies. Besides providing instant views of different aspects of the Earth, many of the images can now be studied as time-series, revealing vital information on how changes in land use, occupation or natural events are affecting both individual and aggregated bio-systems. The data were originally dispersed worldwide into various archives of the *Landsat* partners, and a major effort is being applied to return all the data to a single site with homogeneous formats and data management. While the challenges with the recent data are largely technical ones stemming from the massive volumes concerned, handling the data from the earliest missions (also the most valuable because of the longer time-bases that they represent) has been fraught with problems of obsolescent or obsolete tape-reading hardware and unknown data formats. *Moral: preserve essential documentation and hardware as well as the data!*

4.1.3 The 'McMoon' Data Hackers

The first photographic images of the Earth from space, taken by the *Apollo* Lunar Orbiter in its mission to image the surface of the Moon at very high resolution, were developed on site by the satellite, scanned in strips and beamed back to Earth. However, computing hardware of the day could not cope with the volumes and full resolution of the downloaded data, so the tapes onto which the data had to be recorded and stored were only partially read and analysed at the time, and then put onto a back-burner somewhere (out of sight and consequently out of mind). The tapes were found about 40 years later in a decommissioned McDonald's restaurant in Silicon Valley. Reading them correctly required all the ingenuity of a team of computer 'hackers' (plus finding, acquiring and refurbishing the necessary tape-readers), before the original images could be studied properly in all their glorious detail (wikipedia.org/wiki/Lunar_Orbiter_Image_Recovery_Project). *Moral: Once you eliminate the impossible, whatever remains, no matter how improbable, must be the truth (Arthur Conan Doyle).*

4.1.4 Digitizing Historical Stream-Flow Data

Levels of water in Cape Town's reservoirs seemed to be falling faster than could be due to increasing city populations alone. The cause was discovered by digitizing 74 years' worth of handwritten stream-flow rates measured for the streams that fed the reservoirs. The region had been reforested some decades earlier, and the *non-native* trees that had been planted in their place were taking up more water than did the ones that had been cleared. *Moral: Nature knows best.*

4.1.5 GODAR: Facts from the Oceans

The huge diversity of specimens and features in Nature requires the use of statistics when assessing properties that embrace large amounts of data. However, historical data also come in a wide diversity of forms and formats, as GODAR (the international Global Oceanographic Data Archaeology and Rescue project; nodc.noaa.gov/General/NODC-dataexch/NODC-godar.htm) discovered when assessing and digitizing numerous historical oceanic measurements. The objective was to complement more recent measurements and to fill in gaps in the information records. Sea levels, salinity, temperatures and plankton profiles are particularly important for understanding decadal variations and climate change, and need to be made available from numerous places on the globe and in long series in order to support statistics that are adequately sound. Proposed, designed and pursued persistently until established, the GODAR campaign dealt with paper manuals, microfiche and early electronic media of non-contemporary readability, as well as computer-ready files; the researchers applied suitable quality controls and were able to triple the number of ocean profile data types that had been recorded before the project commenced in 1992 [6].

More recently a large addition of science-ready, hourly electronic data were also rescued [7]. The achievements of the new science snowballed, stimulating new investment into the project such that more data could be ingested, more data made available for oceanographic and climate research, and other associated projects were undertaken. *Moral: keeping the vision simple will pay handsome dividends.*

4.1.6 Regrouping and Reworking Climate Data

NOAA (USA) has digitized important sets of historical data from both routine sources and from more unexpected ones, ranging from ships' logs (see 4.1.7) and army forts to regional climate centres and national weather bureaux, giving priority to samples whose media were deteriorating. The data types extend from daily weather maps, ionospheric and balloon measurements and marine logs to the more conventional but still analogue-only weather recordings [8]. The earliest data were even extracted from diaries of the Founding Fathers and contemporary colonists [8]. Variability information could also be extracted by digitizing proxies like ice cores, tree rings and lake sediments, again often fighting the deleterious effects of time on the physical media. The substantial advances in our understanding of the Earth's climate are in no small way due to their derivation from such a broad cross section of disciplines, because our forecasts of the future state of our climate hinge sensitively upon our ability to reconstruct past climates. *Moral: push the boundaries wide.*

4.1.7 Logs from Nineteenth-Century Ship Voyages

The logs of US Navy and Coastguard ships recorded during voyages dating back well over 150 years have proved a particularly valuable source of information for climate research. The data are extensive, so the international partnership, Old Weather (oldweather.org), set up an online service that uses volunteer assistance ('citizen scientists') to digitize many of the handwritten measurements of the weather (air pressure and temperature, wind speed, precipitation, etc.) from the logs. The digitized versions of the data will constitute an enormous new resource for modelling and understanding both historical and future weather patterns worldwide. Sharing such data with the public through 'crowdsourcing' also inspires a whole portfolio of powerful new ideas and new concepts. *Moral: do what you can with what you've got.*

4.2 Reworking Heritage Data for Trans-Disciplinary Research

Science itself evolves and not infrequently some measurements made for project **A** may be recognized much later as critically valuable for project **B** even though the latter was very far removed from the purposes of the original observers. The next two examples illustrate that well.

4.2.1 An Unusual Data Chase

This example offers a snapshot of an epidemiologist at work in a Spanish archive. As he turned through paper records dating from the seventeenth century, he sniffed each one. He was following the progress of an epidemic of cholera. How? Vinegar was used by the post offices of the time to disinfect correspondence coming from infected sites, and its smell never goes away (private communication).

4.2.2 Deriving Stratospheric Ozone Concentrations via Astronomical Data

Observations of the sky made from the ground unavoidably include signatures from the Earth's atmosphere. While that property has driven astronomers to launch telescopes into space or build them on high mountains in order to get above the atmosphere, it was put to advantage in the case of stratospheric ozone. That upper atmospheric layer protects every bio-system on Earth from destruction by the Sun's harmful UV radiation. Routine monitoring of ozone concentrations was pioneered in England and established in Switzerland in 1926, but—as is always the case for pioneering programmes—the early data were less plentiful or reliable than later ones, so confirmation of the results from those start-up years rests upon identifying subsidiary sources of information; the strengths of the ozone features are also very sensitive to meteorological conditions. UV spectra of hot stars recorded at different observatory sites have recently been recognized as an excellent, yet completely untapped, source for investigating the UV ozone bands. Some of the quality observations date back to the 1930s and earlier. The data are photographic and required specialist handling, but the project proved its point [9, 10] and demonstrated the inescapable added value of reusing data in a trans-disciplinary mode. Even so, more data are still required in order to bring the project to a robust conclusion, and attempts to digitize all the necessary plates have had to wait for technical reasons.

4.2.3 Heritage Data of Pluto Enhance the Best of Modern Measurements

The astrometric satellite Gaia is currently determining highly precise positions for over a billion stars and the brighter solar system objects. Despite its intrinsic precision, the resulting ephemeris of objects like Pluto can still be improved upon by including other astrometric measurements taken many decades ago. There is a suitable set of such measurements (on photographic plates) in the plate archive of

Pulkovo Observatory, Russia. The limitation to previous attempts at accurate astrometry from those plates lay in the inferiority of the digitizing equipment rather than in the plates themselves. Plates taken in 1930–1960 have recently been digitized with the high-precision scanner of the Royal Observatory of Belgium. The project, still ongoing, has been described by Khrutskaya et al. [11].

4.3   Serendipitous Discoveries of 'Lost' Data

4.3.1   Found! the Original Ozone Monitoring Records

During the course of the project mentioned in 4.2.2 above, the author visited a small museum in Oxford (UK) which housed some of the original equipment and records acquired during the early days of global ozone monitoring. Inside a cupboard, bound in red tape, were bundles of the first records from the 'master' instrument in Oxford. Concurrently, ozone experts at ETH Zurich (Switzerland) were revising the analyses of all the early measurements of ozone concentrations as reported by the several stations in Europe that were part of the monitoring network, but had been unable to find any tabulations of the ones from Oxford. It took one email, a phone call, and a visit from the Swiss researchers to work on digitizing and re-analysing the records, and the missing information was restored, much to the satisfaction of everyone involved [12].

4.3.2   Dusty Medical Data Recently Given a New Lease of Life

Clinical trials carried out 40–50 years ago to investigate relationships, if any, between heart disease and (a) cholesterol levels (newscenter.lbl.gov/2011/01/04/cholesterol-heart-disease) and (b) amounts of unsaturated fat in the diet (cbc.ca/news/health/the-case-of-the-missing-data-1.1412420) have both now received due recognition of their value to medical research, though they only received scepticism at the time of first analysis and publication. Controversial results are not popular when they go against the grain of current popular thinking, but fortunately in both cases the data were not destroyed and were eventually recovered and reworked. The lapses in time since the data were acquired and the original studies that were carried out have actually enriched both sets of data by extending their time-bases to complement modern data significantly, though it was tough on the individual researchers at the time who pioneered the studies and lost prestige through their work rather than gaining it.

4.4   The Human Factor

The new knowledge that is gained can be either esoteric or general, but may sometimes be dramatic, and critical to life. All kinds are essential. Actually, knowing the rhythms of flooding versus drought years in semi-desert regions can enable farmers to plan ahead so that bad years do not find them unprepared. Accurate weather forecasting is another clear instance in which human suffering and loss of life can be reduced. Studying long-term patterns of El Nino and La Nina offers a means for predicting and understanding not only the behaviour of populations of fish and their spawning patterns, but also what humans can expect, and prepare for, by way of extreme weather.

5   Preserving and Transforming Heritage Data

5.1   The Storage Environment

If heritage data are to be made publicly available (as they need to be for full dissemination), and their information content transformed into files and formats that are fully digestible by modern computers, there clearly needs to be agreement on, and application of, basic standards and protocols. The last situation that one wants to encounter is the requirement for *re*-recovery of data that were inadequately 'rescued' by well-meant but misplaced enthusiasm. Those basic standards and protocols will of course depend in detail on the nature, as well as the extent and condition, of the data in question. Old books may need to be protected from UV light, which is damaging to some old materials; photographic films and plates should not be handled by non-experts in case the emulsion is contaminated by fingerprints or scratches, and if washing is required in order to remove grime or accumulated bloom and dirt, then that is certainly a matter for consulting the expert. Old (or even relatively new) 7-track or 9-track magnetic tapes require a tape-reader which, while not quite obsolete yet, is certainly obsolescent in many places. Designers and allocators of storage spaces also need to consider carefully the requirements of the materials to be stored, and the likely damage that can result if ignored. Setting aside a special building with its own lock and key may sound sensible for storing a valuable set of photographic plates, but when the designated storage rooms are upstairs in a warm climate and the building has a tin roof...?

5.2   Specialized Hardware

Tempting though it may be to discard ill-maintained and often bulky pieces of hardware that are no longer used, like readers for magnetic tapes, punch paper tape or punched cards, the rescue of pre-born-digital heritage data may still depend on that equipment, so a clearing house needs to be established where seekers can connect with hardware that has been declared redundant. Forecasting when that time arrives is not possible and requires the designation of a backup repository of such items somewhere, which can help out if needed.

5.3 Digitizing Technologies and Methodologies

The variety of types of heritage data is broad, extending from paper and paper-like products such as books and journals via charts, pro formas and records that may be printed, photographed or handwritten, to microfiche, early magnetic tapes and their later cousins like DAT tapes and floppy disks (do you remember floppy disks?) Each new invention was touted at the time as *the* salvation for data storage for modern science, and can be matched by a correspondingly wide span of interpreting and translating technologies and tools. Offices have become revolutionized by the flatbed scanner, but it is important to respect what it was designed to do, and to limit its use to appropriate kinds of materials. A flatbed scanner will create a digital reproduction of a sample or image, but often there is inbuilt software that 'beautifies' the original in some supposedly beneficial way, for instance by softening sharp edges between light and dark even when those same sharp edges are actually an essential feature of the original document. The common flatbed scanner employs a bright light which a photographic emulsion will scatter somewhat mercilessly, so a scanned image of a photograph will emerge with modifications to its fundamental photometric characteristics and which could be fatal in a *quantitative* use of the digital version. OCR is a powerful, and in some circumstances highly efficient, tool for converting a lot of printed figures or texts to tables of numbers or documents, but it may need patient and careful 'training' and is unlikely to work at all for handwritten characters. Often the choice of which technology to use is unfortunately bent by financial constraints, and the effort involved becomes wasted when the method followed is inappropriate.

5.4 Engaging Volunteers

It is often suggested that one engage volunteers to do the legwork involved in retrieving heritage data and carrying out what needs to be done to 'rescue' a heritage archive. However, parts of the activity are likely to need skills that are above and beyond what a typical 'citizen scientist' will probably be able to offer, while maintaining adequate standards, as well as training and supervising individuals, may take up more time than attempting to do most of the job oneself, leaving rather little that can actually be left safely in the hands of unskilled personnel. One problem with recruiting volunteers involves an interesting crisis of *perception*: the status that is accorded to a particular task seems to match the status of the person doing it, so if some task is declared to be suitable for unskilled or untrained effort, then it must hold rather little value in the eyes of the tenured scientific community.

6 Where Do Heritage Data Fit into the Current Data Landscape?

6.1 Current Trends

Scientists used to complain that they could never get enough data. Nowadays the boot is on the other foot: aided by advances in capacity and speed of computers that were unimagined until relatively recently, the modern scientist is in serious danger of drowning in the floods of data which his own ingenuity has now made possible. Coping with the coming tsunamis, in part to justify the expense and resources consumed by the new equipment and computational support, absorbs a great many of the available resources but offers chances of career enhancement to the early-career scientist who is thus able to be right on the front line where new science is being uncovered, and can also become fluent in modern digital skills. Clearly (or so the argument goes), there is a pressing need to justify the capital expense of the new equipment that has been built and the running costs that are absorbing at least the resources allocated to it, and probably more besides. That justification will need to feature the scientific results, and even though amply planned and simulated well in advance, reality always manages to come up with unforeseen challenges that demand immediate attention until solved. It is the way of life, particularly that of a deeply involved researcher: an inelastic system of people and projects has somehow to incorporate, and make space for, additional loads that are bound to stretch it relentlessly. New assistance may be brought in, but probably not enough to release those already newly engaged to devote much time to continuing with yesterday's project work at anything like the same rate before the interruption. Something else has to give, and so often the first thing to get stood down is that attention to the science's heritage.

There is no intended criticism implied in the above. What appears to have occurred is a very common situation in life and describes an action that any of us is likely to choose, or have wished upon us, at any time and in any place. The omission (if one can be identified) lies rather with a system of management of people, projects and sub-systems that do not recognize the need for a champion of each feature that is its declared interests. In astronomy, for example, the workhorse detector for many decades was the photographic plate, and each observatory equipped with a research-quality telescope and a large enough output of observations appointed a Plate Librarian to classify new materials and to catalogue, track and recall the holdings just as though they were books in a conventional library.

The fairly abrupt switch to digital detectors and ready access to the corresponding data online rendered the support of a Plate Librarian rather suddenly unnecessary; those who did not have transferable IT skills found themselves made redundant, and unfortunately much of the wisdom that had been invested in the plate catalogues and holdings was also sent to a back-burner—indefinitely.

6.2 Reversing the Trends

6.2.1 Encouraging the Data Seekers: The *Guidelines for Data Rescue*

As this chapter has tried to describe, there is a somewhat negative connotation associated with the use of heritage data in modern research. The situation enjoys a temporary reversal when a new result of sufficient significance is revealed because old data that held critical information could be included in the research in question, but does not acquire enough new momentum to retain a lasting interest in heritage data. Efforts are currently underway to improve the lot of this aspect of research through a number of parallel initiatives. Very often, those who struggle to work with the kinds of nearly lost data described above feel—and are—isolated, unpopular and unsupported. An important basic step, therefore, is to create a network which can link together as many relevant individual groups and efforts as possible. Another step is to offer assistance that will help pilot the project through some of the uncertain routes and choices that can end in setbacks if not navigated optimally. To that end, a set of *Guidelines for Data Rescue* (rd-alliance.org/guidelines-data-rescue-0) offers pointers and advice accumulated through the experiences of a wide assortment of projects, practitioners and performances. In data rescue work, each project is an individual and the *Guidelines* cannot expect to address all of the problems presented by each one, but they offer a range of suggestions to provide constructive solutions and feasible alternatives. Moreover, by being available online they are not set in stone and can be updated when new ideas or experiences are reported.

6.2.2 Building a Global Connectivity

The trends that we are currently perceiving in matters concerning heritage data do have positive aspects, but they need to be nurtured energetically if they are to grow and prosper. One key activity is building communications that link together various data-rescue groups. It is also vital to maintain strong, positive advertising about the activity of data rescue, by giving talks, publishing (research, views, reviews and reports; e.g. [13]) and holding workshops. A significant forward stride was achieved by the Boulder Workshop—cementing links, reaching out and debating future plans (codata.org/blog/2016/09/10/the-data-at-risk-task-group-dar-tg-in-expansive-mood).

As hinted at the start of this chapter, those who champion the rescue of heritage data may have to contend with the stigma of being unpopular and old-fashioned as well as with all the other factors that sharpen competitions for funding, staffing and other research priorities. Countering those attitudes will be aided by involving a sufficiently critical mass of people, by making it easier to access historic data for incorporation alongside modern ones, and by producing an expanding dossier of Case Studies. Planning an effective network itself takes time and effort, but pays back handsome dividends.

7 Heritage Data and Future Science

While it may be a truism to state that scientific research is only as complete as the range of its data, it is an unassailable truth that heritage data can extend that completeness uniquely by augmenting the ranges of time, place and kind. In the very great majority of cases, the richness of heritage data is therefore to be found in their complementarity to modern data, a way of adding 'instant history' to our research resources. Examples in which the data themselves constitute the prime or sole input for research will mostly be found in the domains of history (as 4.2.1). Even so (to follow up that example), understanding the agents that have caused epidemics of disease in the past can be critical in formulating the most effective defences in the future. In other words, scientific data do not stand in isolation. Always, for some purpose and in some form, data from all ages enrich their science; they provide key pieces of information and enable improved understanding. Nearly-lost heritage data—so often uncatalogued, uncharted, somewhat derided, left behind by technology and in increasing risk of being discarded through lack of use stemming from lack of access—are every bit part of the data landscape as are the overgrown tracks, the fallen trees, the derelict barns and the dried-up stream beds of our natural world. Each data set contributes its own story to the assembly of information that furnishes the resources for research, and each must be respected, curated and made accessible for all. The quality of our future science hinges critically upon how that accessibility is managed.

The topics to which heritage data seem—at present—mostly likely to be able to add significant riches involve our changing planet, and involve any of the myriad aspects which lead to current evidence about it. There is an obvious time span to consider; the most influential records are those that date from the years or decades which predate the onset of the changes that seem to be occurring, even accelerating, now. But science too is evolving, as shown by the example in 4.2.2, and it is not possible to predict what relevance yesterday's data may have to tomorrow's explorations. However, we can, and must, learn from past experiences. The atmospheric scientists in example 4.2.2 who guided the determination of historic ozone concentrations quickly recognized the latent potential of the unfamiliar source of data, and asked the ultimate question, 'Why were these data not made available to us years ago?' Why, indeed. As in every other science, astronomers have their hands full to overflowing coping with reducing and interpreting astronomical data for astronomy.

Data management tasks are surely something for the data scientists, tasks for which they are both trained and qualified to handle advantageously. Who has not heard the adage, *Behind every successful scientific researcher is an expert data manager*? It applies to scientific data of all forms, all disciplines and all epochs. The dual tasks of 'getting' and 'sharing' are what have made the world of science what it is today, but if designed and treated as partnerships, those dual tasks can lead seamlessly to more widespread information, rapidly enhanced knowledge, and more appropriate responses to situations, humanitarian or otherwise. The resources, training and facilities are all in place; it only requires rearranging them, reassigning tasks, and designing the vision.

Working with that model will ensure that no data will again risk becoming forgotten, and it will place the twenty-first century on record as being the start of a new, richer, brighter era in scientific research.

Acknowledgements It is a pleasure to acknowledge the generosity of the Dominion Astrophysical Observatory in Victoria (NRC, Canada) for my continuing privileges as a Visiting Worker, and the use of its astronomical facilities in any of the specific projects mentioned in this chapter. I would like to acknowledge once again the financial support received from CODATA International in order to attend numerous meetings in which issues of Data Rescue were raised, and to repeat our gratitude to the RDA, Elsevier Publishing and the Alfred P. Sloan Foundation for specific grants to run the Boulder Workshop in 2016. I am also grateful to colleagues worldwide who share my interest in raising awareness of all heritage data that risk getting forgotten, and whose support in such endeavours is a continuing source of inspiration.

References

1. Güntsch, A., Fichtmüller, D., Kirchhoff, A., & Berendsohn, W. G. (2012). Efficient rescue of threatened biodiversity data using reBiND-workflows. *Plant Biosystems, 146*(4), 752–755.
2. Vittoz, P., Bodin, J., Ungricht, S., et al. (2008). One century of vegetation change on Isla Persa, a nunatak in the Bernina massif in the Swiss Alps. *Journal of Vegetation Science, 19,* 671–680.
3. Walther, G.-R., Beißner, S., & Burga, C. A. (2005). Trends in the upward shift of alpine plants. *Journal of Vegetation Science, 16,* 541–548.
4. Guicking, D., Fiedler, W., Leuther, C., et al. (2004). Morphometrics of the punk-footed shearwater (*Puffinus creatopus*): Influence of sex and breeding site. *Journal of Ornithology, 145,* 64–68.
5. Tschöpe, O., Wallschläger, D., Burkart, M., & Tielbörger, K. (2011). Managing open habitats by wild ungulates browsing in North-Eastern Germany. *Applied Vegetation Science, 14,* 200–209.
6. Levitus, S. (2012). The UNESCO-IOC-IODE global oceanographic data archeology and rescue (GODAR) project and the world ocean database projects. *Data Science Journal, 11,* 1–26.
7. Caldwell, P. (2012). Tide gauge data rescue. In L. Duranti & E. Shaffe (Eds.), *The memory of the world in the digital age: Digitization and preservation* (pp. 134–149). Paris: UNESCO.
8. NOAA. (2008). *Climate Database Modernization Program*, NOAA, 2008, Annual Report.

9. Griffin, R. E. M. (2005). The detection and measurement of telluric ozone from stellar spectra. *Publications of the Astronomical Society of the Pacific, 117,* 885–894.

10. Griffin, R. E. M. (2006). Detection and measurement of total ozone from stellar spectra: Paper 2. Historic data from 1935–42. *Atmospheric Chemistry and Physics, 6,* 2231–2240.

11. Khrutskaya, E. V., De Cuyper, J. -P., Kalinin, S. I., Berezhnoy, A. A., de Decker, G. (2013). *Positions of Pluto extracted from digitized Pulkovo photographic plates taken in 1930–1960.* arXiv:1310.7502K.

12. Vogler, C., Brönnimann, S., Staehelin, J., & Griffin, R. E. M. (2007). Dobson total ozone series of Oxford: Reevaluation and applications. *Journal of Geophysical Research (JGR), 112* (D20), 2156–2202.

13. Griffin, R. E. M. (2017). www.nature.com/news/rescue-old-data-before-it-s-too-late-1.21993.

Big Data and Predictive Analytics: A Facilitator for Talent Management

Neetu Jain and Maitri

1 Introduction

Arguably the most practical tool and greatest potential for organizational management is the emergence of predictive analytics. —Fitz-enz and Mattox II [3]

Big data analytics is the process of collecting, organizing, and analyzing large sets of data (called *big data*) to discover patterns and other useful information. "Big data" means the volume, variety, and velocity of data that resides in most companies. Big data is large volume of data from a number of sources which is getting generated at good volume, variety and velocity [5]. Waves of Internet like Internet of things, Internet of knowledge, Internet of people, and Internet of everything have turned this world into a global village. Frequent interactions and interface among people and organizations have resulted in emergence of big volume of data. Statistics suggests that the users of various social networking sites such as Facebook, Twitter, online Websites, LinkedIn, electronic and mobile commerce companies, and blogs generate large amount of data. Electronic footprints of cyber interactions and all online activities lead to data generation. Technological operations of smart cards, ATM, and many other sources have also generated extremely large data sets. This data can be analyzed by applying business intelligence to reveal behavioral interactions, patterns, and trends. Just collecting data has less business significance. Generating insights and value from this data helps in talent management decisions. Analytics brings value to the data.

N. Jain (✉)
Indian Institute of Public Administration, New Delhi, India
e-mail: drneetujain76@gmail.com

Maitri
Management Education and Research Institute, New Delhi, India
e-mail: maitri.sawarn@gmail.com

© Springer Nature Singapore Pte Ltd. 2018
U. M. Munshi and N. Verma (eds.), *Data Science Landscape*,
Studies in Big Data 38, https://doi.org/10.1007/978-981-10-7515-5_14

1.1 Predictive Analytics

Capturing, sorting, and customizing the data as per the business requirements is the biggest challenge for a business. If it is done in an appropriate manner, it can help in confident projections, decision making, and strategy formulation. Predictive analytics is a real-time approach to data mining, keeping in view user convenience and business prediction.

2 Objective

The objective of the paper is to throw light on the concept of big data and its role in management of talent. Consulting and research organizations provide different human resource services as per the competitive requirements to the organizations. This paper aims at discussing the significance of big data for HR professionals and use of this data to create value for the organizations in the most cost-effective manner. Google uses the word people operation for all its human resource services as they are being managed mathematically. Managing humanitarian aspect can be analytical and mathematical as it helps to overcome many manual challenges.

3 Role of Big Data in HR

Analytics present an immense opportunity to help organizations understand what they do not yet know… By identifying trends and patterns, HR professionals and management teams can make better strategic decisions about the workforce challenges that they may soon face. —Huselid [4]

Experts discuss applying big data to marketing and consumer businesses; there is an even greater opportunity to apply big data to human resources. It is called as talent analytics—now rechristened as people analytics. HR's use of big data is in its infancy [2].

3.1 Predictive HR Analytics

Predictive HR analytics is systematic application of predictive modeling using inferential statistics to existing HR or people related data in order to inform judgments about possible causal factors driving key HR-related performance indicators. Put simply, sophisticated statistics and quantitative analysis techniques are used by scientists to predict things. Talent analytics is utilization of available big data from human resource management perspective. It focuses on optimizing people operation strategy. Workforce analytics is designed to leverage business intelligence technology for human resource decision making. Development of workforce analytics to get talent insights is dire need for organizations in order to optimize performance and cost. Alignment of human capital management policies to business strategies supported by effective data can act as a business differentiator. Ability to retrieve right information about the available talent is critical for organizations. Talent acquisition for the available talent pipeline with good financial viability is required to be done by the organizations. Bersin [1] outline the importance of using predictive analytics to help organizations predict and understand the performance of a person based on available historical data. According to 15th Annual Global CEO Survey carried out by PWC, more than 80% of US CEOs say that they need critical talent-related insights to make business decisions, but only a small percentage of US CEOs actually receive relevant information. This analytics can help deliver the promise of the right workforce, the right hires, reduced turnover, a robust talent pipeline, an engaged team, and the achievement of financial and operational goals [6]. Research quoted by Forbes estimates that more than half of companies sampled (over 60%) are investing in big data and predictive analytic tools for use in guiding human resources decisions. To navigate the competitive advantage for business viability, human resource transition and transformation is must. Human resource leaders have started using advanced analytics for talent decisions. Frequency mapping, regression analysis, trend analysis help in concrete visualization about human capital management decisions. HR is gravitated toward visualization based on predictive and advanced analytics.

4 Three Dimensions of Big Data in HR

Data from multiple sources provides insight to employers about prospective employees which eventually helps in customizations of human resource policy. Presence of people who are also prospective applicants on social and professional networking sites acts as a window into employees' personal and professional lives. Tracking and analysis of this data provides complete picture about the prospective employees' attitude, behavior, personality, and aptitude which will help organizations to get insights into matching the job profile with them. Robust workforce analytics can be a game changer.

The business world is being transformed by the volume, speed, and availability of data. This has made people and performance even more critical for business sustainability. All the research on talent analytics has focused on three key aspects:

- **Technology**
  What type of systems, processes, and infrastructure can drive data and talent analytics? What platforms are being used and how can we use systems such as Oracle, Hadoop, and the like to develop a coherent data strategy?
- **Techniques**
  How an approach to talent analytics can be developed? How data can be defined, stored, and shared? How should data be analyzed? This aspect covers everything from defining employee turnover to predicting patterns of employee behavior.

– **Talent**
   What type of talent can deliver the required analytics capability? Should new talent pool be developed and, if so, what types? How can scarce talent be recruited to resource this growing capability need?

5 Application of Predictive Analytics in Human Capital Management

All managerial and operational functions of human resource will get influenced by the inclusion of big data analytics for talent management. Following points throw light on application of talent analytics in human capital management.

1. Data is disparate which can be integrated automatically for operational workforce planning. Gap analysis of finance department, labor budgets, and HR department labor cost can be done with operational workforce planning.
2. Employee surveys will help in understanding accurate manpower requirement and projections. Market intelligence will give strategic insights to the organizations for strategic human resource planning.
3. Manpower surveys can also be conducted to evaluate impact of existing HR strategy. Outcome of this data will help to explore future avenues according to business requirements.
4. The jobs of data scientist, data analyst, and data managers are the lucrative ones in the market. These people are required to be hired for organizations, so responsibility of HR has got increased to draft job analysis, job description, and specifications for the job related to big data professionals.
5. Analytical and smart manpower acquisition decision can be predicted with the help of analytics. Sensible hiring can only be made after accurate manpower planning.
6. Analyzing employee information from different sources can help HR to perceive accurately intention of employees to stay or to quit. Employee satisfaction factors can also be studied and improved by tools like employee satisfaction surveys, team assessments, and social media, exit and stay interviews. Further appropriate policy can be customized for better satisfaction and retention. Xerox used big data analytics to cut its attrition rate at call centers by 20%.
7. Predictive analytics can be used to validate selection process as per the specific requirements. Company can integrate recruitment strategy along with social networking to gather more information about applicants on social and professional front. Accurate person job match or person organization fit can be analyzed by posted pictures, tweets, etc.

8. Big data analytics can also help employers to identify and acknowledge top performers, along with workers who may be struggling in their positions. Investing in talent management software can assist HR professionals in gathering and analyzing the data they need to evaluate individual performance levels.

9. Benchmarking practices of industries can be explored and standardized as per the organizations and industry requirements.

10. Predictive analytics can be utilized on 360 data to see key driver competencies for business. Performance management can be done accurately overcoming all biases. Key performance indicators can be evaluated according to key responsibility areas by utilizing precise analysis.

11. Competency mapping can be based on these data to identify core competence of employees, thereby imparting training and development accordingly. It will lead to achieving high performance.

6 Conclusion

This study has explored changing face of human resource management. Human resource has a long way to go. Technological challenges have redefined the human resource management. Digital reorganizations have brought paradigm shift in human capital management functions. HR is required to embrace this change for survival and sustainability. HR professionals are required to prove themselves not just as people champion but also change enabler, strategic business partner, and market differentiator. Programs, practices, performance, and productivity of people can be managed, enhanced, and enriched with the technological transformation. HR professionals can crunch numbers. Fusion of human capital management and predictive analytics can do wonder for strategic human resource management area. Competitive challenges of current perspective can be explored and addressed efficiently with the help of inclusion of big data into human capital management. Indeed, human resource professionals can lead the organizations by becoming strategic technical business partners.

References

1. Bersin, J. (2012). The HR Measurement Framework, Bersin and Associates Research Report, November.
2. CIPD. (2016). Talent analytics and big data—the challenge for HR (online). Accessed April 30, 2016. http://www.cipd.co.uk/hr-resources/research/talent-analytics-big-data.aspx.
3. Fitz-enz, J., & Mattox, J. R., II. (2014). *Predictive analytics for human resources*. New Jersey: Wiley.

4. Huselid, M. A., & Becker, B.E. (2005). Improving human resources analytical literacy: Lessons from moneyball. In D. Ulrich, M. losey, & S. Meisinger (Eds.), Future of Human resource management. Wiley: New York.
5. Mcafee, A., & Brynjolfsson, E. (2012). Big data: the management revolution. *Harvard Business Review, 90*(10), 60–68.
6. Pricewaterhousecoopers. (2012). *Key trends in human capital in 2012: A global perspective*, [online]. London: PWC. www.pwc.com/gx/en/hr-management-services/key-trends-in-human-capital-2012-a-global-perspective.jhtml.

Privacy Preserving Data Mining Techniques for Hiding Sensitive Data: A Step Towards Open Data

Durga Toshniwal

1 Introduction and Motivation

The advancement in technology as well as the dropping in the prices of storage has made both the storage as well as analysis of humongous amounts of data possible. Further, the possibilities of parallelization has also reduced the time and effort needed for this previously cumbersome task. Due to these advancements, several new use cases for data mining have evolved. For instance, consider the area of medicine, various medical institutes can now mine their data collectively and gain useful insights in a very short period of time using parallel mining techniques. There can be similar use cases in different domains like marketing, weather forecasting. The only problem that still prevails is that in most cases when collaborative data mining is performed data privacy is lost. The owner of the data may wish to hide some sensitive information while using the benefits of mining as well. So collaborative data mining may allow better planning, better decision making, and better business strategies, but privacy is one of major issues to be handled here. Privacy preserving data mining (PPDM) is the branch of data mining which aims at developing techniques that can be used to extract knowledge from sensitive data without any loss of its privacy. As proposed in [1], PPDM also concerns analysis of data mining techniques to determine how they affect the privacy of data. It aims at deriving useful aggregate patterns from the data while keeping the individual sensitive data elements hidden. The owner of the data must be able to engage in collaborative data mining while keeping his/her data hidden from the miner as well as all the other parties involved in the mining task.

All the techniques used for PPDM suffer from a major challenge of achieving a good balance between data utility and data privacy. In most cases, data is perturbed

D. Toshniwal (✉)
Department of Computer Science & Engineering, Indian Institute of Technology Roorkee,
Roorkee 247667, India
e-mail: durgatoshniwal@gmail.com; durgafec@iitr.ac.in

© Springer Nature Singapore Pte Ltd. 2018
U. M. Munshi and N. Verma (eds.), *Data Science Landscape*,
Studies in Big Data 38, https://doi.org/10.1007/978-981-10-7515-5_15

or modified in order to hide the sensitive information. This hiding of data decreases the quality of data thus affecting the data mining results obtained from mining such data. There is always a trade-off between data privacy and the accuracy of mining results obtained in case of these techniques. The advent of new technologies like cloud computing, distributed computing has increased the need for better and newer PPDM algorithms. There are two broad approaches to PPDM—Introducing privacy preservation in the data or embedding privacy preservation in the data mining algorithm. Various algorithms pertaining to both of these approaches have been proposed.

In this report, we have organized the existing privacy preserving data mining techniques along two main lines. One approach towards PPDM is publishing privacy preserving data. In this case, privacy is introduced into the data set itself by performing various operations on it. Data mining techniques can then be applied on this data without any loss of privacy. The other approach towards PPDM is when privacy preserving methodologies are embedded into the data mining process itself. These kinds of techniques generally cater to one particular data mining problem like classification or association rule mining and introduce privacy preservation into the algorithm itself.

There are various factors that act as an impetus to the development of PPDM techniques and its study as an interesting area of research:

- The realization of the potential benefits that can be achieved through collaborative mining is in itself the greatest motivation toward privacy preserving data mining techniques. The loss of sensitive information is a threat that may lead to the data owner not engaging in collaborative mining. On the other hand, such data mining tasks can facilitate the extraction of very valuable information from the data. For instance, consider a scenario where one of the wholesaler wishes to mine the data of his retail sellers collectively to determine the demand patterns accurately. Now consider there is one retail seller who sells many of the wholesalers' products and also the wholesaler enjoys a monopoly at his/her store. In such a situation, leaking this information will make the retail seller's dependence on the wholesaler visible to the wholesaler may lead to the wholesaler taking advantage of his vulnerability and increasing the prices. In such a situation, the retail seller needs some way in which he can share the data he has with the wholesaler for mining while preserving the hidden information.
- Mining on data while preserving its privacy may seem like an easy task but in reality it involves a lot of issues. The data utility may drastically decrease while introducing privacy into the mining process thus leading to inaccurate results. So PPDM techniques are needed which introduce privacy preservation while maintaining the utility of the data.
- There is a need of PPDM techniques that introduce privacy preservation into the conventional data mining processes without increasing their computational complexity much.

- More and more new scenarios that require the use of PPDM techniques are coming up with new technologies. For instance, there is need for PPDM techniques that are specially adapted to suit environments like cloud.

Thus, the need of better PPDM techniques, dearth of good PPDM algorithms for scenarios like distributed databases, cloud environments, and the growing number of businesses investing in data mining in order to facilitate better decision-making capabilities serves as a good motivation toward the study of PPDM.

2 Different Types of Techniques for PPDM

Though a variety of categorizations of PPDM techniques have already been proposed but based on our study, we have categorized them as shown in Fig. 1.

As evident from Fig. 1, the studied techniques can be categorized under two broad categories:

- Generic Privacy Preserving Data Mining Techniques
- Specific Privacy Preserving Data Mining Techniques

Generic approaches are the approaches that introduce privacy preservation into the data in such a way that the transformed data can be used as input to perform any data mining task. In this report, three generic techniques have been discussed. Details of these techniques are described in Sect. 3. The specific techniques are the techniques that apply to only some particular data mining task. Privacy preservation in this case is embedded into the particular data mining algorithm allowing the mining to be performed without harming privacy. In this report, we discuss three specific PPDM techniques. All the three techniques apply to different data mining problems as shown in the Fig. 1.

Details of these techniques are discussed in Sect. 4.

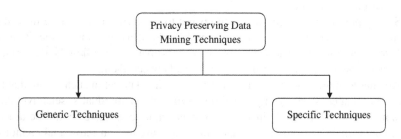

Fig. 1 Categorization of the privacy preserving techniques

3 Generic PPDM Techniques

In this section, firstly some generic PPDM techniques are described that transform the data set in such a way that all the sensitive information in the data is hidden. Any of the data mining tasks can be performed on this data without any breach of privacy.

The first technique described here proposed in [2] perturbs the data set in order to preserve privacy. In this technique, called slicing, data is partitioned into buckets and columns are formed by grouping highly correlated attributes together. The column values may be generalized to preserve privacy if required. In order to make reidentification impossible and ensuring the safety of sensitive data from getting revealed, values in a column are randomly shuffled.

The limitations in the existing techniques like generalization and discretization for anonymizing data motivated the development of this technique.

- Generalization suffers from the curse of dimensionality. For instance, in case of k-anonymity generalization, it is required that at least k records should be very close to each other. In case of high-dimensional data achieving this constraint requires large amounts of generalization thus decreasing the utility of data.
- When all attributes are generalized separately the correlation between attributes is lost. As a result, the data after generalization has very poor utility at times and thus unsuitable for mining.
- Discretization provides better data utility than generalization but it does not prevent membership disclosure because in discretization the values for quasi-identifying attributes are not modified. So it may be possible to determine whether the data contains a particular entity's record or not.
- Discretization requires a clear division into pseudo-identifying and sensitive attributes.

Slicing preserves the linking within each column whereas breaks the linking between different columns. Attributes having less association between them are more identifying in the sense that their combinations occur infrequently. By breaking associations between uncorrelated attributes slicing reduces the chances of their use for identification.

Slicing preserves more utility as compared to generalization because instead of putting the same generalized value for each tuple, it assigns one value from the multiset of all possible values. This preserves information regarding the distribution of values in spite of breaking the associations between them.

Another technique described in [3] transforms the data set in such a way that the data set satisfies k-anonymity thus preserving the individual's sensitive data. K-anonymity is the property in which for every individual there should be at least k indistinguishable records in the data, thus ensuring that an individual cannot be correctly identified in a k-anonymized data set with a probability greater than $1/k$. The approaches used for k-anonymity can be categorized on the basis of whether k-anonymity is enforced at the time of data publishing or at the time of generating results for data mining. Two possible approaches to ensure k-anonymity are:

- *Anonymize and Mine*—Generally a two-step process.

  - Pros: A clear separation between data protection and mining. Data mining is safe as it is performed on the anonymized table. The task of mining the data can be carried on by anyone.
  - Cons: Not suitable for stream data. The cost of anonymizing the data is more than the usefulness as the data is mined only once. Mining is carried on with less complete and generalized data so the quality of inferences is reduced. The anonymization algorithms lead to information loss.

- *Mine and Anonymize*—Generally a one-step process.

  - Pros: Prevents the distortion of data and preserves its completeness and usefulness as anonymization is performed on the mining results. The quality of data mining results is good.
  - Cons: The data mining can be carried on only by the data holder.

The one-step process requires a change in the data mining algorithm to incorporate the creation of anonymity whereas the two-step process allows the use of the existing data mining algorithms as it is.

In terms of computation time, the one-step process works better than the two-step process.

The technique described in [3] performs k-anonymization on the Web search query logs released by Web engines. This particular technique will provide good results in data mining techniques like clustering, classification, categorization. However, this technique suffers with some shortcomings. The technique does not take into account the number of queries pertaining to a single user and therefore the information loss is higher for users with less number of queries. Also the solution compares queries only from syntactical point of view, semantic aspects are unexplored.

In [4], privacy is preserved by adding noise to the private attributes of the data. In this technique, noise is added to all the attributes of the data so as to preserve the data privacy. The noise addition technique varies with the type of attribute. Noise is added to both categorical and numerical attributes. The process is carried out in two steps:

1. First, all the sensitive class attributes are perturbed.
2. Second, all the non-class attributes are also subjected to noise so as to avoid reidentification of records through them. This also prevents these attributes from getting revealed.

Adding noise to numerical attributes is simple as compared to adding noise to categorical attributes. Here, a clustering-based technique is described to efficiently add noise to categorical attributes while maintaining an acceptable level of data utility. The resulting data has good utility. When decision tree learning was performed on the perturbed data, the resulting decision tree was very close to the decision tree obtained from the original data. But the technique requires a lot of computation and is complex.

4 Specific PPDM Techniques

In this section, we discuss some of the recent PPDM techniques that aim at performing some particular data mining task while preserving privacy. The techniques use algorithms that perform mining while hiding the sensitive data.

More and more organizations today are moving their data to the cloud and also using the computational resources physically present on the cloud because of cloud features like elasticity, economy of scale, and many more. In such scenarios, data mining processes are to be performed over large data sets and that too while hiding the sensitive data as data from different organizations can be stored on a single cloud. PPDM algorithms are needed which can be used to effectively mine this data while keeping every organization's data private. The first technique described here performs classification on database stored on cloud while keeping its privacy intact. The classifier used in this technique is K-nearest neighbor (KNN) [5]. The technique makes use of cryptographic algorithms for secure exchange of data. The parties collaborate together and globally classify records with a help of a trusted third party. Secure multiparty computation (SMC) is used by the parties to calculate the class of a test record [6]. The technique gives very high level of privacy. The points, the distance values as well as the local class value data—all three are secure throughout the computation. But as cryptographic techniques are involved the biggest issue here is the theft of key.

Another technique described in this section creates perturbed data sets from the original datasets which appear to be very different from the original dataset but impart the same meaning conceptually when used for the task of decision tree classification. This technique is based on complementation operation of set theory [7]. The benefits of this technique are that the data can be perturbed at storage time itself and also mining results are very accurate. But the space complexity of this technique is very high. Moreover, security is completely broken if an attacker gets access to all the perturbed datasets.

5 Some Cloud-Based Techniques

In the present section, we discuss some of the existing research works on PPDM which utilize cloud technology.

Anonymization using MapReduce framework on cloud is a parallelized data hiding approach as proposed in [8]. The scheme handles the issue of scalability in case of anonymization techniques when dealing with large voluminous datasets.

A scalable two-phase top-down specialization approach is presented in [8]. A scheme for privacy preserving data utilization in hybrid cloud is proposed in [9]. A practical hybrid architecture is proposed in which a private cloud is introduced as an access point interface between the data owner and the public cloud. Under this architecture, a data utilization system is provided to achieve both exact keyword search and fined-grained access control over encrypted data. The approach realizes an outsourcing cryptographic access control mechanism and further reduces the computational cost at the user end. Another privacy preserving association rule mining in cloud computing environment is proposed in [10]. The scheme considers the scenario where data owner encrypts the data and stores it in the cloud. For mining association rules from such encrypted data, three solutions are provided to protect data privacy during association rule mining. The solutions are built upon distributed ElGamal crypto-system and achieve item privacy, transaction privacy, and database privacy, respectively.

6 Conclusion

In the present survey, we discuss various recent PPDM techniques. We describe the techniques by categorizing them along two main dimensions. Out of the total six techniques discussed, three are generic and apply to all kinds of data mining tasks and the rest three apply to specific data mining scenarios. In this study, we make the following observations. Among the generic techniques, slicing performs very well. It maintains a very good balance between privacy and utility though it is a bit expensive computationally. The second generic technique is k-anonymization of query logs and this does not achieve good balance between privacy and utility. With increasing privacy, the utility of the data decreases drastically. The third technique which involves transforming the actual data to privacy preserving data through noise addition, performs well in providing privacy while maintaining a considerable amount of utility but is relatively computationally intensive. The biggest problem with the specific techniques is that they are specific to a particular data mining task. There is thus scope of developing PPDM algorithms that provide improved level of privacy while preserving data utility to a considerable extent, which can be used for more than one data mining task and are also not very expensive and complex in computation.

References

1. Verykios, V. S., Bertino, E., Fovino, I. N., Provenza, L. P., Saygin, Y., & Theodoridis, Y. (2004). State-of-the art in privacy preserving data mining. *ACM SIGMOD Record, 3*(1), 50–57.

2. Li, T., Li, N., Zhang, J., & Molloy, I. (2012). Slicing: A new approach for privacy preserving data publishing. *IEEE Transactions on Knowledge and Data Engineering, 24*(3), 561–574.
3. Navaro-Arribas, G., Torra, V., Erola, A., & Castella-Roca, J. (2011). User k-anonymity for privacy preserving data mining of query logs. In *Information processing and management* (pp. 476–487). Elsevier.
4. Islam, M. Z., & Brankovic, L. (2011). Privacy preserving data mining: A noise addition framework using a novel clustering technique. In *Knowledge-based systems* (pp. 1214–1223). Elsevier: Amsterdam.
5. Ibrahim, A., Jin, H., Yassin, A. A., & Zou, D. (2012). Towards privacy preserving mining over distributed cloud databases. In *International conference on cloud and green computing*.
6. Clifton, C., Kantarcioglu, M., Vaidya, X., Lin, J., & Zhu, M. Y. (2002). Tools for privacy preserving distributed data mining. *SIGKDD Explorations Newsletter, 4*(2), 28–34.
7. Fong, P. K., & Weber-Jahnke, J. H. (2012). Privacy preserving decision tree learning using unrealized data sets. *IEEE Transactions on Knowledge and Data Engineering, 24*(2), 353–364.
8. Zhang, X., Yang, L. T., Liu, C., & Chen, J. (2014). A scalable two-phase top-down specialization approach for data anonymization using mapreduce on cloud. *IEEE Transactions on Parallel and Distributed Systems, 25*(2), 363–373.
9. Li, J., Li, J., Chen, X., Liu, Z., & Jia, C. (2014). Privacy-preserving data utilization in hybrid clouds. *Future Generation Computer Systems, 30*, 98–106.
10. Yi, X., Rao, F. Y., Bertino, E., & Bouguettaya, A. (2015). Privacy-preserving association rule mining in cloud computing. In *Proceedings of the 10th ACM Symposium on Information, Computer and Communications Security* (pp. 439–450).

Role of Credible Data in Economic Decision Making

Geethanjali Nataraj and Ashwani Bishnoi

1 Introduction

In an environment of uncertainty, different stakeholder's namely public policy makers and business executives rely on maximum possible information for sound decision making. Timely availability of relevant and accurate data helps in making appropriate decisions in an uncertain environment. Decision making is the process of identifying alternative courses of action and selecting an appropriate alternative in a given situation. Prudential decisions are the key to effective policy making. But decision making is largely dependent on available information. For instance, relevant information about the financial performance of a company helps the investors to make right choices about the security selection and similarly correct information related to employment in countries gives a path for the next course of monetary policy actions and other policies. For financial development, the financial reporting is serving the cause of effective and efficient resource allocation for financial institutions.[1,2] Credible, independent data are critical not only for research but also for a proper functioning of democracy. It is also observed that the government's choice not to collect or release data often reflects vested interests.[3] India has slowly

---

[1] FASB (1999). International standard setting: a vision for the future. Norwalk.

[2] IASB (2008). Exposure Draft on an improved Conceptual Framework for Financial Reporting: The Objective of Financial Reporting and Qualitative Characteristics of Decision-useful Financial Reporting Information. London.

[3] Dean Spears and Diane Coffey (October 14, 2015). Measuring well-being, The Indian Express.

G. Nataraj (✉) · A. Bishnoi
IIPA, New Delhi, India
e-mail: geethanjali_n@yahoo.com

G. Nataraj · A. Bishnoi
NIT, Kurukshetra, India

© Springer Nature Singapore Pte Ltd. 2018
U. M. Munshi and N. Verma (eds.), *Data Science Landscape*,
Studies in Big Data 38, https://doi.org/10.1007/978-981-10-7515-5_16

declined from a world leader in the availability of survey data to a place where meaningful statistics simply are not available.

1.1 Importance of Incredible Data

Economic decision making has a different nature starting from the corporate house to public sectors utilities. In corporate sector, economic decision making refers to the process of making business decisions involving money. In public sector, the decision making involves effective policy making. Overtime, it has been realized that relevance and reliability of information are two important ingredients for decision making. Relevant data information possesses two characteristics of time-liness and predictability. Delay in information compilation is worthless for policy makers. Regarding predictive capacity, it means that before economic decision makers opt for one alternative instead of the other, they must be sure about the reward of policy.

Business executives make choices of investment among various alternatives and strategies based on real-time data. This phenomenon has fueled the growth of big data to enable the corporate houses to make better, smarter, real-time, data-driven decisions. Businesses take the help of data by controlling the new opportunities while managing the new risks.[4] Moreover, the concept of big data has evolved not only in terms of more volume but also in terms of more sources giving space for effective decision making with relatively reliable and consistent databases.

In the recent past, the global community has utilized huge amounts of resources in amassing information, analyzing it, and reaching decisions on the problems confronting them. On the development front, the goals of sustainable development have been well announced amid the partial achievements of Millennium Development Goals (MDGs). Herein the countries have considered the need for high quality, reliable and multi-dimensional data and statistics. At the UN Statistical Division, there are 230 indicators which are used to gauge the progress on the SDGs, across more than 190 countries. The data supports both—effective decision making in line with investment and proper monitoring and a guide for future course of strategies.

It was revealed that availability of data and proper analysis puts forth the scope for better prediction and optimum usage of resources.[5] In the field of agriculture, it was announced that the advancement of agriculture is the key to feed the world's rapidly expanding population.[6] But the advancement rests on proper extraction of

---

[4]Big data Changing the way businesses compete and operate, http://www.ey.com/Publication/vwLUAssets/EY_-_Big_data:_changing_the_way_businesses_operate/%24FILE/EY-Insights-on-GRC-Big-data.pdf.

[5]Barton D, Court D. Making Advanced Analytics Work for You. Harv Bus Rev; 2012.

[6]Gilpin L. How big data is going to help feed nine billion people by 2050. Tech Republic; 2014.

available resources for farming and utilization in the best possible manner which requires the proper information gathered through data. In the health sector, it was reported that efficiently analyzing various forms of healthcare data over a period of time can answer many of the impending healthcare problems.[7] Siemens and Long have understood the role of big data for shaping the future of higher education.[8] In the advanced countries, researchers have started to go in depth to explore every possible information. As an instance, Picciano (2012) has explored the students' progress in online courses as well as identified the difficult subjects to understand. In this regard, the students' activities like login/logout information, number of mouse clicks and number of page views, time spent on a page, and contents posted by student have been tracked.[9] It is also hailed that the integration of information and services for citizens facilitates internal governance as well as transparency of the government through external access.[10]

The quality of data sets and the inference drawn from such data sets are increasingly becoming more critical, and organizations need to build quality and monitoring functions and parameters for big data. For example, correcting a data error can be much more costly than getting the data right the first time—and getting the data wrong can be catastrophic and much more costly to the organization if not corrected. Moreover, the periodicity of data collection is equally important. Herein the availability of baseline data is important for the socio-demographic and economic indicators. Baseline data can be used to determine the groups' equivalence before the program began or to "match" different groups They are also important for determining whether there has been a change overtime and how large this change is (i.e., the effect size).

1.2 National Initiatives for Improving Statistics

The policy makers have understood the need for a sustainable ecosystem that brings healthy partnership among various stakeholders such as industry players, government, and academia. Moreover, the big data is gaining its ground for wide-ranging applications in different spheres of economy. In this context, the Government of India has committed to strengthen the database with due security. The launch of the

[7]Jensen P.B., Jensen L.J., Brunak S. Mining electronic health records: toward better research applications and clinical care. Nat Rev Genet. 2012; 13: 395–405.

[8]Siemens G, Long P. Penetrating the fog: analytics in learning and education. EDUCAUSE Review. 2011; 46(5): 30–2.

[9]Picciano A.G. The evolution of big data and learning analytics in American higher education. J Async Learn Network. 2012; 16(3): 9–20.

[10]IDC Big Data and Business Analytics Conference 2013; 2013 Nov; Abu Dhabi, United Arab Emirates. Available from: http://idc-cema.com/eng/events/54217-idc-big-data-andbusiness-analytics-conference-2013/7-overview.

National Data Sharing and Accessibility Policy (NDSAP)[11] in 2012 is the landmark initiative to supplement the big data. National Informatics Centre (NIC) under the Ministry of Communication and Information Technology is responsible for providing the technology-based data management through Data Portal of India.[12] Digital India is an initiative to integrate the government departments and the people of India and aims to promote electronic transactions and reducing paperwork. The NIC has created the Open Government Data portal, data. gov.in. Currently, 85 government ministries, departments, and agencies have contributed more than 12,000 data sets across segments such as population census, water and sanitation, health and family welfare, transportation and agriculture to data.gov.in.[13] The major concern is that the digital India initiative is to be systematically linked in such a way that there is automatic updating of data at different locations in the country to a central pool.

Ministry of Statistics and Programme Implementation (MOSPI), a merger of two departments, statistics and program implementation, is working since 1999 to compile the national level statistics. It has two wings—NSO[14] and program implementation.[15] Keeping in view the importance of data, MOSPI had released a strategic plan for period 2011–16 whereby the data collection processes had been dealt at the national level. The prominent strategies include the cooperation and coordination of all the stakeholders such as Government Agencies, Research Institutions, Individual Researchers, Private Corporate Sector, Trade/Industry Associations, Data Providers, and Media. In order to improve the quality of data, measures such as improving the collection problems or processing problems, training of field staff, documentation and improved data processing practices, regular interaction between data source agencies and data users, conducting methodological studies to improve the sample design have been considered. In order to have the proper flow of administrative data from state to the center, a regular interaction between the statistical advisers posted in central administrative department, State DESs, Line Department in States and State Capital DDGs of NSSO (FOD) has to be ascertained. This strategic guide map of MOSPI is to be realized with proactive engagements of different stakeholders through better technology

---

[11]The Policy aims at the promotion of a technology-based culture of data management as well as data sharing and access.

[12]http://www.archive.india.gov.in/spotlight/spotlight.php#tab=tab-1.

[13]Ajit Kumar Roy, April 24, 2017, Big Data and Data Science Initiative in India, Biostat Biometrics Open Acc J 1(2): BBOAJ.MS.ID.555560 (2017).

[14]The National Statistical Organization (NSO) consists of the Central Statistics Office (CSO), the Computer Centre and the National Sample Survey Office (NSSO), is the nodal agency for planning and facilitating the integrated development of statistical system in the country, and to lay down norms and standards in the field of official statistics.

[15]The Programme Implementation Wing (PI Wing) is responsible for monitoring and evaluation of major projects and programmes of implementing Member of Parliament Local Area Development Scheme (MPLADS).

management mainly the banking networks, postal networks, smart phones, retail networks, etc.[16]

Other measures include the technical guidance related to methodological issues for sample surveys, financial support to the states for taking up surveys, to promote development of satellite accounts of various important sectors like health, environment, education, trade, construction have been envisaged in the vision plan. Regarding the release of report annually with maximum one year lag, the scale of institutional capabilities is to be strengthened in terms of opening more sub-offices in different parts of India and conducting the survey of macro-indicators at regular and continuous level.

The limited availability of data has three shortcomings: poor assessment of strategic gap, difficulty in comparison across countries and regions, and problems related to the measurement of real effect of public policies. In order to make effective economic decision making, this chapter makes an attempt to present the current status of data sources, underlines the gap for present databases, and discusses the procedures for bridging the gap related to data. The study also aims to explore the best and innovative approaches for the databases management in India. The rest of the chapter is organized as follows. Section 2 describes the current economic environment and the use of data in today's times for various purposes including calculation of national output, compilation of industry data, data related to agriculture, external sector, government finance, money and banking statistics, employment, economic development, and infrastructure. Section 3 details the requirements of different types of data for India's development agenda. Section 4 puts forth the measures for data improvement/credible and reliable data, and Sect. 5 provides recommendations on how to enhance the database in the country ensuring easy accessibility of credit data for policy making. Finally, Sect. 6 the conclusion highlights the missing points in Indian data with respect to various sectors also provides a road map for credible data collection in the country.

2 Assessment of the Current Environment

In the era of information technology, policy makers should have the ability to acquire relevant information on various subjects simultaneously. Policy makers need access to vast data for policy making.

---

[16]MOSPI (GOI), Document Strategic Plan 2011–16 of Ministry of Statistics and Programme Implementation, released by Ministry, http://mospi.nic.in/sites/default/files/main_menu/strategic_plans/Strategic%20Plan_MOSPI_10022011_Final.pdf, accessed on June 2017.

2.1   Macroeconomic Aggregates

The fields where the database is reasonably good include the information pertaining to macroeconomic indicators mainly national income, investment, savings, inflation, trade, and foreign capital flows. But in a few areas, the country lags far behind than its counterparts. The broader categories for databases are: administrative units, agriculture, banks and financial institutions, civil supplies and consumer affairs, companies, cooperatives, crime and law, demographics, economy, education, environment and pollution, foreign trade, forest and wildlife, geographical data, health, housing, industries, insurance, labor and workforce, market forecast, media, metrological data, mines and mineral, petroleum, power, power, rural and urban areas, social and welfare schemes, telecommunication, tourism, transport, etc. India lacks substantially in major policy-relevant indicators either due to much lagged period availability or unavailability of time series, or no availability. Most of the data that have been collected or produced are not made readily available even in aggregated form. The major limitations attached to the access of data are in terms of data files from raw data stored in inaccessible databases, huge dependency on ad hoc sources—newspaper articles, experts' opinions, interest groups, or lobbyists. Where do policy makers go for information? Obviously, policy makers in India face a wide range of issues and so potentially need information about several major topics, but this discussion will focus on major economic aggregates and development types of data.

2.1.1   National Output

India has a strong foundation in the compilation of data related to macroeconomic indicators namely national output, value addition, investment expenditure, savings, inflation, trade, at aggregate, disaggregate level, and regional level. Central Statistical Organization is the sole body compiling the national income data and same is facilitated by Handbook of Statistics on Indian Economy a publication of Reserve Bank of India. The components of gross domestic product data are available at full length since the 1960s. Even the frequency of data is available at quarterly level. The national level demand scenario is captured through index of industrial production, PMI,[17] and gap between demand and supply is reflected in indices related to inflation. The trade and foreign investment statistics are available at full length. National Account Statistics (NAS) presents the data related to gross capital formation and capital stock but again it is confined to the broader sectors, the disaggregate level data still carries a lot of scope which needs to be tapped. Though industrial investment is available through Annual Survey of Industries, investment in agriculture sector for recent periods is not available to get compiled. Regarding the macro-indicators former Chief Statistician of CSO, Pronab Sen mentioned that

---

[17]Purchasing managers index.

"IIP has limitations to be considered as a representative of aggregate growth in manufacturing. But it is still useful and the only measure available for the micro-picture for a broad sense of what is happening to different segments. We continue to use IIP for estimating capital formation and construction because there is nothing else available as a leading indicator." It was also added that there is disconnect between formal and informal sectors particularly when 35–55% of India's output may come from the informal sector. He also stated that India considers the "formal sector data as a proxy" for the growth of informal economy.[18] This position clearly speaks of data gap related to disaggregate level compilation.

2.1.2 Industry

Industrial statistics is mainly compiled by Ministry of Statistics and Programme Implementation through Annual Survey of Industries. The limitation is that ASI mainly covers the data pertaining to the organized sector. NSSO surveys the unregistered sector at the national level with greater time lag and the frequency of once in five years thereby rendering no availability of data in intervening years. Even the questions related to reliable estimates of value of inputs, output, employment, and value added, even at 2-digit level of industrial classification were raised by economists.[19] Periodical collection and publication of statistics for the unorganized sector is the most promising field for database. Ministry of Micro, Small and Medium Enterprises (GOI) maintains the information related to the expenditure incurred for various schemes, viz. technology up-gradation, credit support, marketing promotion schemes, skill development, infrastructure development, and physical performances through MSMEs benefitted, trainees benefitted in terms of numbers, etc. But again the data related to nature of utilization of expenses under various schemes by individual firms and nature of benefits for MSME firms (individually) are considered as critical data needs. Moreover, India is experiencing the era of start-ups and major thrust can be to explore the provision for detailed databases from the very beginning.

2.1.3 Agriculture

For data relating to agriculture statistics, the primary sources of data are Directorate of Economics and Statistics, Department of Agriculture and Cooperation which compile disaggregate level data for Indian agriculture. Commission for Agriculture Cost and Pricing (CACP) surveys present the data for cost of cultivation across

---

[18]How Do We Resolve the Confusing Puzzles Surrounding the Latest GDP Data? The Wire, 02/03/2017.

[19]M.R. Saluja, Industrial Statistics in India: Sources, Limitations and Data Gaps, Economic and Political Weekly, Vol. 39, No. 48 (Nov. 27–Dec. 3, 2004), pp. 5167–5177.

crops. Agriculture Statistics at Glance (ASAG) is a rich source of data related to agriculture, however, all the sources lag behind in terms of compilation of data related to varietal developments at aggregate level as well as crop wise level. Also, the country lacks behind in data related to technological advancements at regional level or state-wise. The biggest challenge in India for agriculture sector is to address the inefficient supply chain management. So far India has no specific database capturing the data pertaining to supply chain management. Other gaps in agriculture data seem—head wise spending of research and development, agricultural nitrous oxide and methane emissions; technology development, inventions of machines and software, etc.

2.1.4   External Sector

For data relating to external statistics, the primary sources of data are export–import database of Ministry of Commerce and Industry and Directorate General of Commerce Intelligence and Statistics (DGCI&S), Government of India (GOI). The trade performance data is available at aggregate and sub-sectors level. However, certain limitations are there related to services trade data. Presently, India compiles the services data for standard classification like travel, transport, and telecommunications. However, information by partner country, by categories and sub-categories of services as prescribed in the negotiating list W/120, and by mode of delivery of services is not available from the RBI data. In this context, the impact assessment of service agreements across regions cannot be carried out.[20] The Department of Industrial Policy and Promotion compiles the data for capital flows mainly inward. However, the frequency of data at disaggregate level is limited and the sector-specific data for foreign investment reaching to state destinations is not available in India.

2.1.5   Government Finance

For data relating to state finances, the primary sources are the Department of Economic Affairs and Ministry of Finance, Government of India. The time series compilation of data at aggregate level and region level is available from the Handbook of Statistics on State Finance, a publication of Reserve Bank of India. In fact, the data for public spending and receipt components for various heads are available for India with respect to time series. But the data gap related to utilization of public funding at disaggregate level is quite missing in the Indian context. Also, such data needs to be compiled at regional level to see the efficiency of funds utilization across Indian states.

---

[20]Press Information Bureau, Government of India, 30-January-2017, Developing an Institutional Framework for Capturing Services Trade Data—Challenges and Opportunities.

2.1.6 Money and Banking Statistics

The monetary indicators data is available through India's Central bank where components of money, interest rates, reserve requirements, foreign exchange reserves, etc. are presented at full length. The frequency of these statistics ranges from the weekly to fortnightly to monthly and annually. With the wider availability of rich information related to financial indicators, the previous RBI governor could make appropriate decisions and accordingly the problems related to inflation and exchange rate instability have been managed. On the front of financial system, statistics is widely available in India. Starting from monetary policy instruments to bank statistics to stock market trading, vast information is available. The key data sources are Reserve Bank of India and Bombay stock exchange and National stock exchange. A serious limitation appears that banks have data statistics related to the credit expansion for aggregate level. Bank credit data is available for limited sectors such as mining and quarrying, food processing, sugar, tea, beverages, textiles, leather products, wood products, paper products, chemical products, iron and steel, electronics, gems and jewelry, power, roads. With growing NPAs in the system, the firm-specific bank credit data needs to be compiled. Such data will help in understanding the role of bank credit in very sector and also can be helpful in understanding the pass-through effect of monetary policy changes mainly interest rates. Since the country has launched the marginal cost-based lending rates (MCLR) hence a data related to mapping of deposit rates and lending rates for various sectors needs to be compiled. Gaps in banking statistics include point-of-sale terminals for rural and urban areas, borrowers, and depositors from commercial banks; statistics related to crop insurance deliveries by the cooperative banks, etc.

2.1.7 Employment

In the recent past, it is witnessed that trends in employment are one of the most important inputs in setting monetary policy responses as has happened in the USA. But India has access to time series data only in the organized sector that constitutes a tiny part of overall work force of India and there is no data on the unorganized sector. Annual Survey of Industry compiles the firm-specific data, but with a large lag that makes it virtually useless for monetary policy purposes.[21] Employment data in India is mainly compiled by National Sample Survey Organization. However, the latest data for employment in India is available for year the 2011–12 as compiled by 68th round of NSSO Employment and Unemployment Situation in India. The other data source for employment is Labour Bureau, Ministry of Labour and Employment. It is observed that Employment and Unemployment Surveys (EUS) of National Sample Survey Organization (NSSO) provides comprehensive

---

[21]https://mostlyeconomics.wordpress.com/2009/07/06/limitations-in-indian-economics-statistics-system/.

information with an interval of five years. However, Labour Bureau has been conducting surveys annually since 2010. Hence there is duplication of data compilation resulting in higher costs and instead, the exercise can be carried out at alternate years by both the agencies. Furthermore, the reliability of employment data has been a matter of concern. For example, worker participation rate was estimated to be 38% for male persons and 25% for women in rural India as per NSSO (EUS) for 1999–2000 while these rates were estimated to be 61 and 58, respectively, on the basis of the Time Use Survey (TUS) for 1998–99 conducted by CSO.[22] Here, the coverage of employment nature and types needs to be well defined.[23] Gaps in employment data include unemployment with various education levels, labor force with different education levels, and composition of employment like self-employed wages and salaried persons, the skills and training of the workforce, more detailed data on the supply and demand for workers in specific industries and/or occupations, and more detailed workforce projections.

2.2   Economic Development Indicators

Development deals with improvement in living standard of the people. In this regard, fiscal policies need to be designed keeping in view the present status of development across regions of India. The soundness of fiscal policy depends on availability of information related to demographic characteristics, level of education, and health facilities across nearly every sector of society. However, the barriers to access and use of development indicators' data are found more acute in developing economies. Even the challenge related to availability of high quality is the key factor for poor planning and assessment of development interventions.[24]

2.2.1   Education

For data relating to education, the primary sources of data in India are District Information System for Education (DISE) and UGC. These two organizations compile the statistics related to education indicators for schools and higher education, respectively. Education attains significant weightage in the computation of economic development. In education, the prominent indicators are enrollment rates and literacy rates. For enrollment rate, the problem related to over or

---

[22]Hirway, I., Jose, S. (2011). Understanding women's work using time use statistics: The Case of India, Feminist Economics, 17:4, 67–92

[23]T.S. Papola 2014, An Assessment of the Labour Statistics System in India, International Labour Organization Country Office for India New Delhi.

[24]Demographic Data for Development Decision making Case Studies from Ethiopia and Uganda, Population Reference Bureau Assessment Team Angela Akol, Jason Bremner, Assefa Hailemariam, Grace Nagendi, Negash Teklu, and Charles Teller.

underestimation is attached whereas literacy rate covers the poor measurement of education. Even literacy data is available with lag periods. However, information related to indicators namely social characteristics of the students such as religion, economic particulars of the family (parental income, education, occupation, etc.), net attendance rates and unit costs for different educational levels among social groups are not available.[25] Gaps in education data are information relating to adult and career education and more detailed enrollment and graduation data, compensation to education staff, current education expenditure for all levels, viz. primary, secondary and tertiary, educational attainment, population with at least Master's or equivalent degree, educational attainment, at least Bachelor's or equivalent, trained teachers in all education levels, etc.

2.2.2 Health Data

Health indicators are key to economic development. The major sources are Ministry of Statistics and Programme Implementation (MOSPI), GOI; National Health Mission for rural and urban areas, Rural Health Statistics under open Government data platform. WHO India also compiles the statistics related to health parameters such as fertility rates, live births, number of deaths, life expectancy, prevalence of serious diseases like HIV AIDS, malaria, tuberculosis; mortality rates, utilization of health services, viz. contraceptive prevalence, antenatal care, births attended by skilled health personal, etc. But health statistics seem to have suffered due to nonalignment with standard definitions. Infant mortality rate is the most important indicator for health performance but affected by inappropriate sample size, and having no coverage with respect to sex, rural, and urban areas (for smaller states and Union Territories).[26] Gaps in health data are non-availability of data for number of doctors with respect to specializations, number of beds, cost of accessing the health facilities, private sector penetration in the health sector, loads on government hospitals (per 100,000 population), etc. It was observed that health and nutrition data in India released by the Rapid Survey of Children are often available for policy purposes at an aggregate level but not at household level which overlook the inequality and obscure details of household behavior.[27]

---

[25]Draft for discussion, Statistics on Education: Advantages and Limitations for Studies on Social Groups, P. Duraisamy, University of Madras, preliminary draft for discussion at the National Seminar on "Social Group Statistics and Present Statistical System: Emerging Policy Issues, Data Needs and Reforms" jointly organised by Indian Institute of Dalit Studies and Ministry of Statistics and Programme Implementation held at New Delhi during February 23–24, 2012.

[26]Oommen C. Kurian on 11/01/2017, Why It's a Challenge to Make Quick Sense of India's Health Data, The Wiire.

[27]Dean Spears and Diane Coffey (October 14, 2015). Measuring well-being, The Indian Express.

2.2.3 Demographic Data

Three main sources of demographic data in India include Census, Civil Registration System, and Demographic Surveys. Census covers the plausible indicators for demography and includes distribution of population across sex, age, literacy rates, occupational pattern, marital status, religion. The sample registration system (SRS) provides reliable annual data on fertility and mortality at the state and national levels for rural and urban areas separately. Quality of data on fertility has not been very good, but after making adjustments using some demographic techniques, one can use the estimates of fertility at the district level. Despite the compulsory registration of births and deaths in India, not all births and deaths are reported, especially in rural areas due to illiteracy of people and ignorance of the role of registering these vital events.[28] National Family Health Survey (NFHS) that was carried out first in 1992–93 (NFHS-1) covered the data at national and state level for indicators such as fertility, nuptiality, family planning use and demand, unwanted fertility, utilization of Reproductive and Child Health (RCH) services, breast feeding and weaning practices, child nutrition and health, infant and child mortality and knowledge of HIV/AIDS. NFHS-2 was carried out in 1998–99 (NFHS-2) where new topics such as reproductive health, women's autonomy, domestic violence, women's nutrition, anemia and salt iodination were also considered. The rapid household survey under the Reproductive and Child Health project first surveyed 504 districts in India in 1998–99, and second round covered 593 districts as per the 2001 census. The scope of this round was widened to cover birth history that provides fertility and mortality estimates.[29] Again the time lag of data for these indicators is a major limitation for implementing effective health policies.

2.2.4 Data Relating to Social Well-Being

A healthy debate has taken place regarding the per capita income as a sufficient measure reflecting the quality of life and well-being of a person or society. Eminent economists like Joseph Stieglitz and Amartya Sen have suggested the use of broader indicators covering monetary, social, and wellness dimensions. In this regard, broader indicators, viz. home amenities, kitchen facilities, education, hygiene, entertainment, communication, transportation, and health care have been considered to compile the index for well-being. The statistics for these indicators are available at aggregate as well as subregional levels. But the statistics is available with much lagged period and thereby restricts the compilation of index for present time period. Other data gap for social well-being is the lack of data for indicators

---

[28]Divisha S. 3 Main Sources of Demographic Data in India, Sociology Discussion, http://www.sociologydiscussion.com/demography/3-main-sources-of-demographic-data-in-india/3054.

[29]Sources of Demographic Data in India, http://iipsindia.org/pdf/05_b_09cchep2.pdf.

such as coverage of social insurance programs in richest and poorest quintile, benefit incidence of social insurance programs to poorest quintile, coverage of social safety net programs in richest quintile. Some countries effectively use redistribution to reduce inequality, but India is not among them. More importantly, the direct benefit transfer scheme is delivering to the society to reach the benefits to the target but again the poor database of beneficiaries still undermines the scope of scheme.

2.3   Infrastructure

The performance of infrastructure is measured through major indicators such as length of railways and roads, goods transported and passengers carried by railways, container port traffic, passengers carried and freight through air transport, participation of private players in energy generation, water and sanitation facilities, (through investment), mobile cellular subscriptions, improved water source for rural and urban areas, electricity production from different sources such as hydroelectric sources, natural gas sources, nuclear sources. However, this data is available at national level from the source World Development Indicators of World Bank but such statistics are not available at regional level in India. Further, the utilization of energy sources with respect to industry and agriculture is equally important but covered by existing databases. However, the statistics related to time taken by railways and road transport, cost of traveling and freight carriages, number of vehicles passing by particular roads, administrative efficiencies of public transport facilities, etc. needs to be compiled to assess the efficiency of infrastructure resources in the country. There is stringent need for health data compilation related to public–private partnerships. Also, India has started the initiative of digital economy but it requires sound infrastructure for its smooth conduct. Presently, India compiles the national figures related to the exports and imports of ICT services, internet users, secure internet servers, number of smart phone users, etc. But the statistics related to the availability of ICT-related facilities mainly smart phones access, access to computer, internet usage, number of ATM transactions, credit card availability and business carried out under point of sale machines, nature and amount of internet banking, etc. needs to be managed for urban as well as rural areas. Also, these statistics can be explored with respect to demographic characteristics. ICT could help fulfill India's ambition to become a global manufacturing hub and improving productivity in agriculture and the services sector, while boosting access to some basic services among the rural population.[30]

---

[30]*Gemma Corrigan, Attilio Di Batisto (November 05, 2015).* 19 charts that explain India's economic challenge, World Economic Forum. https://www.weforum.org/agenda/2015/11/19-charts-that-explain-indias-economic-challenge/.

The current data environment is not sufficient to meet the requirements of policy makers. It is highlighted that present database has limitations in terms of unavailability, difficulty in obtaining, delay in compilation information, and poor availability of objective data-based analysis. In the existing literature, it is mentioned that lack of sufficient resources allocated to data-related activities and limited implementation of information technology is the main reason for poor database in India. In case of poor access of data, the policy makers rely on far distant proxies to draw the inference about economic phenomenon which may not be useful. In most cases, policy makers have to rely upon special studies prepared by universities or private consultants and often commissioned by private organizations or foundations. The country has a vacuum in the above-stated statistics at the aggregate level, and access of such statistics at regional level is the next promising area. Reports and/or data files that are produced are difficult to use for policy analysis. Most of the agency websites contain limited data and do not incorporate information technology that allows efficient access to the data. Several different and inconsistent regional classification schemes are now in use.

3   India's Development Agenda

The key challenges for the Indian economy in the past decade have been observed in the areas of rural development, urban sustainability, national infrastructure, and human capital and population.[31] On the physical infrastructure front, the priority areas for the country include inter-city rail upgrading, sustainable and secure energy, linkage of rivers for efficient water management. For social infrastructure, the thrust areas are integrated, life-cycle, population scale vision of human capital accumulation which includes addressing the undernutrition and health problems of child. On the education front, key challenges include the quality education for all and education-to-job linkages. In the Global Nutrition report 2016,[32] India stands at 120th position out of 130 countries in prevalence of low weight for height. India is home to over 40 million stunted children under five, more than any other country in the world. India has 282 million illiterate people and 18% of those who went to school were unable to complete secondary school. India has half of the world's 20 cities with the most toxic air.

Ensuring safe, efficient, sound infrastructure cities are the next promising areas to be explored. Entrepreneurs are found discouraged due to red tape and an inefficient justice system along with poor environment of enforcing a contract and resolving insolvency. India scores relatively well in terms of access to finance,

---

[31]Nirupam Bajpai and Jeffrey D. Sachs (July 2011). India's Decade of Development: Looking Back at the Last 10 Years and Looking Forward to the Next 20, Working Papers Series Columbia Global Centers|South Asia, Columbia University.

[32]International Food Policy Research Institute. 2016. Global Nutrition Report 2016: From Promise to Impact: Ending Malnutrition by 2030. Washington, D.C.

however, access to finance remains limited for low-income individuals, especially women. Finance can help poor households optimize severely constrained resources across their lifetime. World Economic Forum (WEF) has identified major potential areas to be pondered over with greater attention. These included reforming the tax code and expanding social protection, reducing red tape and formal entrepreneurship, improve access to sanitation and other basic infrastructure, more equitable and quality education, transparency of public administration and access to finance.[33] Keeping in view the above data gaps and strong demand for development agendas to realize the Sustainable Development Goals, the country needs sound policies and economic decision making which rests on rich information base. It is desirable to improve the periodicity of the survey. It can be conducted on an annual or quinquennial basis like the employment–unemployment surveys.[34] In this context, the country needs to explore every possible avenue to strengthen the databases. The present study considers following points to improve the database system in India.

4   Possible Measures for Data Improvement

The important initiatives for data improvement include the involvement of different stakeholders such as international agencies, research institutions, individual researchers, public, state administration, and private bodies. The measures such as continuous meetings between different departments of ministries, between statistical advisers of states and center, periodic conferences of data users at different levels including Industry/Trade Associations and Establishments, enhancing collaboration between academicians and industry are possible means to enhance the database. Other stakeholders such as Public/citizens can be involved through advertisement in print/electronic media; and Private Bodies Organizations in terms of facilitating research grants, and holding meetings and conferences; and NGOs through seminars/Workshops; and enhancing the training of field inspectors.[35]

Leveraging the technology to expand public and private financial networks in rural areas is an important means to strengthen the data houses. Further, the strengthening of the rural internet network could have implications for reaching the

---

[33]*Gemma Corrigan, Attilio Di Batisto (November 05, 2015)*. 19 charts that explain India's economic challenge, World Economic Forum. https://www.weforum.org/agenda/2015/11/19-charts-that-explain-indias-economic-challenge/.

[34]Duraisamy P. (2012). Statistics on Education: Advantages and Limitations for Studies on Social Groups, Preliminary draft for discussion at the National Seminar on "Social Group Statistics and Present Statistical System: Emerging Policy Issues, Data Needs and Reforms" jointly organised by Indian Institute of Dalit Studies and Ministry of Statistics and Programme Implementation held at New Delhi during February 23–24, 2012.

[35]Document Strategic Plan 2011–16 of Ministry of Statistics and Programme Implementation, released by Ministry, http://mospi.nic.in/sites/default/files/main_menu/strategic_plans/Strategic%20Plan_MOSPI_10022011_Final.pdf, accessed on June 2017.

target groups. In this regard, the initiatives of Government such as Unique Identity (UID) Project, financial inclusion, plan for setting up of 250,000 tele-centers in rural areas to provide public access to computers and online service delivery are the centered measures to help the development agenda of the country. As an instance, UNICEF's "Digital Drum" collects the demographic characteristics of citizens and provides the advisory services related to health, education, and business. Herein the Government can strengthen the digital infrastructure base in remote areas and explore and process information through specialized windows at the regional level.

Data quality has remained a big concern for which the need for a more dynamic, flexible database scheme is the need of the hour. Here, the role of digital data is to be enlarged in terms of development of a system to automatically extract and store the relevant data as and when it is generated.[36] Data storage through cloud technology[37] is the most viable option to store all the industry-related information. The only thing is that the availability of centers should be provided for every sector at any place and at any time at a very reasonable cost. Vertical sectors such as financial services and retail are leading the adoption of mobile and cloud technologies in India. The information related to the point-of-sale (POS) and mobile phones is done in the cloud. The application of cloud technology was utilized by farmers in Sahara, South Africa. The key utilizations are in the fields such as of planting schedules, crop status, harvesting times, and market prices through mobile phones. The cloud-based storage has helped healthcare providers with access to health record of patients remotely through a mobile device. Also, it is mentioned that there must be connection between the data collected from a variety of sources: network and non-network, structured and unstructured. The most important imitative for the data compilation must be to strengthen the integrated database which will be picking the automatic data from various portals such as banks, hospitals, education departments, other Government departments. The cloud technology can serve this purpose more efficiently.

More importantly, the well-defined indicators for various dimensions such as health care, governance, banking, agriculture, industry, development, and infrastructure should be identified.[38] MOSPI has carried out a strategic plan to strengthen the national statistics wherein the involvement of different stakeholders is sought. The frequent meetings and interactions with academicians, researchers, and policy makers can help to find the unique metrics of indicators covering the multi-dimensions of an economy. Also, the discussion among these stakeholders is held to understand the execution plan for data generation.

In the recent past, the implications of big data have been observed in agriculture sector in terms of selection of right agri-inputs, monitoring the soil moisture,

---

[36]Anu Peisker and Soumya Dalai (February 2015), Data Analytics for Rural Development, Indian Journal of Science and Technology.

[37]Cloud-enabled device management helps in user authentication and secure file sharing and syncing of diversified application access.

[38]Peisker, A. and Dalai, S. (February 2015), Data Analytics for Rural Development, Indian Journal of Science and Technology. Vol 8(S4), 50–60, February 2015.

tracking prices of markets, controlling irrigation, finding the right selling point, and getting the right price. Farmers have been extensively utilizing the mobile technology to form the linkage between farm produce and retailing. Herein lays the scope to extract the information through establishing a systematic mechanism of data capture for better decision making.[39] In the field of agriculture, the roles of ICT tools have been identified in accelerating the output of the agricultural sector. The role of wireless sensor network has been identified for monitoring crops, soil moisture, nutrient content and environmental management, security and safety. The remote sensing technologies such as GPS and GIS have been recently utilized by policy makers in assessment of proper crop losses to make the crop insurance scheme a grand success.[40] Here, these technologies can be well managed to compile the information related to agriculture sector.

The success of big data generation depends on the participation and support of both, public and private bodies. Support in terms of finance, standards development, data sharing and access, analytical tools, and technology is the important means to realize the dream of big data. The more utilization of e-Government practices in different administrations can play a vital role to add on the database management in terms of efficiency and periodical reporting of data in the economy (Peisker 2015).

In the energy sector also, India lacks the advantages of a central body for maintaining and disseminating energy data. Herein lays the scope for setting up of a national Energy Information Agency to understand the gaps in energy infrastructure through healthy data, and accordingly design the policies to strengthen the energy infrastructure.[41]

Block chain technology is another advancement to manage the public-sector data while maintaining the security of this information. A block chain comprises blocks and is an encoded digital ledger that is stored on multiple computers in a public or private network.[42] So far, banks, payment-service providers, and insurance companies have shown the highest level of interest and investment in block chain.[43] The same technology can be applied to other sectors to generate the database.

India is considerably behind in data compilation at both the regional level and unit level. All future policies must target these two areas where the huge potential exists for data improvement. Since the country has carried out a mega database project through Aadhar card for every citizen of the country, using the same methodology, the category wise statistics of income, education, demography, health, occupation of the population, etc can be generated and upgraded periodically. The only point is just to set up different windows at state, district, towns, and

---

[39]Big Data for the next green revolution, Updated: January 13, 2017 21:11 IST|Bedanga Bordoloi, Business Line.

[40]CSI Communications. ICT in Agriculture—Indispensable. 2013 Nov.

[41]Rahul Tongia (January 6, 2017). Data management: India needs agency for energy data, OP-ED, Brookings.

[42]Steve Cheng, Matthias Daub, Axel Domeyer, and Martin Lundqvist, February 2017, Using blockchain to improve data management in the public sector, McKinsy and Company.

[43]McKinsey and Company (2015). Beyond the hype: Blockchains in capital markets.

panchayat levels. The windows can be post offices, cooperative banks, societies, institutions, colleges, schools, hospitals, etc. The benefits of the government-run schemes can be attached to this up-gradation. Herein the classification of stakeholders, viz. a salaried person, farmer, worker, businessman, etc is important. Setting up a system for compilation of information at regular intervals within the formal system needs simply linkage. But for the informal system, the country is trying to bring the activities into mainstream through cashless economy and recently announced goods and services tax. As taxes will be levied at production level though collected at the consumer level through appropriate technological advancements. Further, the bar coding of products needs to be made compulsory so that the location of products is confirmed and once the invoice is made compulsory, the details of consumer are linked through mobile number or Aadhar card. Here, technology can help to update the consumption pattern of consumers at the national level. Data gets constantly updated through portals related to consumption pattern. Income status data can also be collected through education departments as the parents of students have to reveal the information and another alternative could be the bank account. For farmers, the central point can be the market committees, authorized dealers dealing with agri-inputs such as seeds and pest controls. Other stakeholders such as retailers and business persons can be targeted through the GST scheme. The only point is that the security of data is utmost important.

5   Recommended Actions

In order to provide the necessary information to policy makers and managers to make informed decisions, the Government needs to work proactively with other stakeholders to establish a Central Data Base. The major suggestions for the improvement of data environment in India can be developing an overall vision for data collection and plan for implementing the initiative. Keeping in view the development agenda of the country, the priority sectors need to be defined. The first preference could be given to the infrastructure database covering sound internet networking, telecommunication, and power sector. The quality and usability of data generated should be ensured through proper monitoring and appropriate training of the field staff. Moreover, the sample survey and its methods need to be upgraded with respect to the need and requirement of different types of data in the system. In order to accomplish the target initially public investment should be made and later on private players may be invited. Also, the exercise of data collection has to be considered as routine exercise and accordingly a certain pie of the annual budget ought to be allocated to support the institutional setup and finance to the state governments for data compilation. One institutional body needs to be set up to cooperate and coordinate the state bodies and manage the database initiative of the country. Also, a campaign at various platforms related to the awareness and usability of data should be encouraged.

In a nutshell, the country can exercise the following processes to strengthen the databases of the Indian economy.

- Defining present economic problems and future vision of the country
- Understanding the need of relevant data and possible indicators
- Improving data collection Process
- Institutional setup for routine compilation of data across all sectors.
- Round table discussions for compilation of broader indicators linking all the sectors of the economy
- Preparing common format for data collection with respect to sectors and standardizing the data
- Planning the execution of data storage: Availability of staff and their training
- Practices related to execution of plan
- Understanding the implementation complexities and revising the execution plans frequently
- Management of data, ensured security
- Public–private partnerships to promote quality data compilation.

6 Conclusion

The importance of credible data has been observed in terms of facilitating sound economic decisions related to business and economy. Data help in designing of effective public policies, strengthening democracy, and better asset choices in financial markets, etc. In India, the fields where the database is reasonably good include the information pertaining to macroeconomic indicators mainly national income, investment, savings, inflation, trade, and foreign capital flows. But in a few of the areas, the country lags far behind than its counterparts. For instance, India lacks substantially in major policy-relevant indicators either due to much lagged period availability or unavailability of time series, or no availability, and very difficult access to information. CSO, NSSO, and RBI are major houses of national macroeconomic indicators. India frequently utilizes the index of industrial production for measuring the demand and supply behavior in the economy but the same has witnessed a number of limitations mainly because of the uncountable informal sector which contributes around 35–55% of GDP. Industry data is available at length in India but information related to the nature of utilization of expenses by MSME under various Government schemes by individual firms and nature of benefits for MSME firms (individually) are considered as critical data needs which are missing in India. Moreover, India is experiencing the era of start-ups and major thrust could be toward exploring the provision for detailed databases from the very beginning. The house of agriculture data is Department of Agriculture and Cooperation under Directorate of Economics and Statistics, and Agriculture Statistics at Glance (ASAG) is the richest source of data. But, the country lacks behind in data related to compilation of technological advancements for crops at regional level or state-wise. The biggest challenge in India for agriculture sector is to address the inefficient supply chain management. So far India has no specific database capturing the data pertaining to supply chain management. Other gaps in agriculture data seem—head wise spending on research and development, agricultural nitrous oxide and methane emissions, technology development, inventions of machines and software, etc. India has good experience in compiling data for external trade and investment. However, the sector-specific data for foreign investment reaching different states is not available in India. Ministry of Finance is the best source of public finance data. But the data gap related to utilization of public funding at disaggregate level is quite missing in the Indian context. Also, such data needs to be compiled at regional level to see the efficiency of funds utilization across Indian states.

The monetary indicators data is available through India's Central bank where components of money, interest rates, reserve requirements; foreign exchange reserves, etc. are presented at full length. However, with growing NPAs problem, strengthened database related to credit expansion for individual firms is the need of hour. The sector-specific credit data is available but with only limited sectors. Other gaps in banking statistics include record of point-of-sale terminals for rural and urban areas, borrowers, and depositors from commercial banks and their credentials; statistics related to crop insurance deliveries by the cooperative banks, etc.

Employment data is a most important indicator for monetary policy actions in the developed world. *But India has access to time series data only in the organized sector that constitutes a tiny part of overall work force and thereby un-accounting the unorganized sector. Annual Survey of Industry compiles the firm-specific data, but with a large lag that makes it virtually useless for monetary policy purposes.* Gaps in employment data also include unemployment with various education levels, labor force with different education levels, and composition of employment like self-employed wages and salaried persons, the skills and training of the workforce, more detailed data on the supply and demand for workers in specific industries and/or occupations, and more detailed workforce projections, etc.

On economic development front, the gaps in education data are information relating to adult and career education and more detailed enrollment and graduation data, compensation to education staff, current education expenditure for all levels, viz. primary, secondary and tertiary, educational attainment, population with at least Master's or equivalent degree, educational attainment at least Bachelor's or equivalent, trained teachers in all education levels, etc. Gaps in health data are non-availability of data for number of doctors with respect to specializations, number of beds, cost of accessing the health facilities, private sector penetration in the health sector, load on government hospitals, etc. Other data gap for social well-being is the lack of data for indicators such as coverage of social insurance programs in richest and poorest quintile, benefit incidence of social insurance programs to poorest quintile, coverage of social safety net programs in richest quintile, etc. Some countries effectively use redistribution to reduce inequality, but India is not among them.

Infrastructure data is available from national and international sources, but the regional compilation of all the predefined indicators of international body is limited in India. Moreover, the country rests on ICT trade for performance of soft infrastructure in India. But keeping in view the recent initiatives of GST and demonetization, the digital infrastructure is of utmost importance for strengthening the soft infrastructure. Information connected to the availability of ICT-related facilities mainly smart phones access, access to computer, internet usage, number of ATM transactions, credit card availability and business carried out under point of sale machines, nature and amount of internet banking, etc. are the promising data requirements.

The government has carried out various initiatives such as setting up of Ministry of Statistics and Programme Implementation, a merger of two departments-statistics and program implementation, launch of National Data Sharing and Accessibility Policy (NDSAP) in 2012, etc. Keeping in view the strong demand for development agendas to realize the Sustainable Development Goals, the country needs sound policies and economic decision making which rests on rich information base. It is desirable to improve the periodicity of the survey. It can be conducted on an annual or quinquennial basis like the employment–unemployment surveys.

The important initiatives required for data improvement include the involvement of different stakeholders such as international agencies, research institutions, individual researchers, public, state administration and private bodies; leveraging the technology to expand public and private financial networks in rural areas is an important means to strengthen the data houses. Data quality has remained a big concern for which the need of hour is a more dynamic, flexible database scheme that needs to be implemented effectively. Here, the role of digital data is to be enlarged in terms of development of a system to automatically extract and store the relevant data as it is generated.

Since the country has carried out a mega database project through Aadhar card for every citizen of the country, using the same methodology, the category wise statistics of income, education, demography, health, occupation of the population, etc can be generated and upgraded periodically. The only point is just to set up different windows at state, district, towns, and panchayat levels and coordination of the same. The windows can be post offices, cooperative banks, societies, institutions, colleges, schools, hospitals, etc. The benefits of schemes can be linked to the up-gradation of data. Setting up a system for compilation of information at regular intervals within the formal system needs simply linkage. But for the informal system, the country is trying to bring the activities into mainstream through cashless economy and recently announced goods and services tax. The taxes will be levied at production level though collected at the consumer level through appropriate technological advancements. Herein, the bar coding of products needs to be made compulsory so that the location of products is confirmed and once the invoice is mandatory, the consumption details of consumer are linked through mobile number or Aadhar card. Here, technology can help to update the consumption pattern of all consumers at the national level. As the country rests on expenditure side to measure the national output, in the same fashion, expenditure pattern can be linked to understand the economic dynamics in the country.

India is considerably behind in data compilation at both the national, regional, and unit level. All future policies must target these three areas where the potential exists to develop data. The possible suggestions are—defining present economic problems and future vision of the country, understanding the need for relevant data and possible indicators, improving data collection processes, institutional setup for routine compilation of data across all sectors, round table discussions for compilation of broader indicators linking all the sectors of an economy, preparing common format for data collection with respect to sectors and standardizing the data, planning the execution of data storage: availability of staff and their training, management of data, ensured security and public private partnerships to promote quality data compilation. Credible, relevant, and latest data is the key to effective policy making in the country, and India has a long way to go in establishing systems for credible and comprehensive data collection and compilation.

Big Data in the International System: Indian, American, and Other Perspectives

Manan Dwivedi

> *The creation, advancement and Diffusion of Knowledge is the Only guardian of True Liberty.*
>
> —James Madison

1 Background

The nomenclature, "Big Data" is a grave misnomer. Its about concentrating a plethora of digitized information in a small microcosm of technology and digitization. Big Data is about the all-pervasive nature and operational content of the Internet systems and the age of connectivity beckons where brevity is the soul of wit in the comprehension, but large labyrinthine algorithms lurk behind the global economy and precision of communication. In the global perspective, the idiom of Simulacra and virtual reality is the order of the day with the level and quantum of digital and electronic exchanges getting augmented between the various nation states. With the advent of the age of WikiLeaks and other whistleblowers, the various governments have augmented their operational ties in maintaining and utilizing data. Various governments in the larger international system and India, too, have an IT policy framework which will form a comparative framework for the research paper.

The question of Big Data enables a scientist to be multidisciplinary in nature, where the individual domains of business, trade and health interests need to be revisited. The domain secure nature of a social scientist akin to the national interest-ordained behavior of a nation state has been delved inside. As an instance, the nom de plume of how social inclusion can be sustained and non-discrimination in the era of civil rights and social legislation can be enumerated in the American quest for big data solution, has got focused attention in the proceeding parts of the paper.

M. Dwivedi (✉)
Faculty of International Relations and International Organizations,
Indian Institute of Public Administration (IIPA), New Delhi, India
e-mail: pollaw@rediffmail.com

© Springer Nature Singapore Pte Ltd. 2018
U. M. Munshi and N. Verma (eds.), *Data Science Landscape*,
Studies in Big Data 38, https://doi.org/10.1007/978-981-10-7515-5_17

2   Setting the Tone for Big Data Quest

What is the meaning of the nomenclature, "Simulacra?" One can safely contend that in the present-day world's convergence-ordained *systeme*, the idiom of a machine-modulated matrix has gradually become the order of the day. As time progresses and the machine completely takes over the functionality of the hand, ear, and the eyes, all leading up to the firming of artificial intelligence (AI), the modulated existence becomes the call of the day. In the same connection, if one were to utilize the idiom of machine modulations, then the Hollywood Cinematic construct, matrix and its trilogy can be cited verbatim as a vast panoramic cinematic experience.

The matrix[1] is an artificially created machine world which can be entered into by traversing a cable where machines are pitted in a battle till finish in an artificially created world lore. The machines and the cyber-byte individuals have become so powerful and all subsuming that they are about to consume the man-world in the physical sense of the term. The rebels are the human beings who are scurrying for cover, and they stay in a liquid-surrounded camouflaged human system amidst the matrix of the machines and the larger cyber world. It all amounts to a Simulacra and a system of virtual reality. The question that arises is that can the world be saved from the machines running riot. The present narrative might appear to be a phantasmagorical overture and fantastical striving, but still the idea of investigating the idiom of Big Data conspicuously cannot be explained away without more than a dabbling inside the matrices of the "make-believe" in order to streamline a complete comprehension of "Big Data."

It was Jean Baudrillard[2] who coined the definition of Simulacra. It can be safely surmised that "Whether or not we live in a world of simulacra, the term is certainly important in light of how we view media. Media theorists, especially Jean Baudrillard, have been intensely concerned with the concept of the simulation in lieu of its interaction with our notion of the real and the original, revealing in this preoccupation media's identity not as a means of communication, but as a means of representation (the work of art as a reflection of something fundamentally "real") (see Footnote 2)". Thus, any representation, be it in the realm of painting, cell phone communication, driving on a four-wheeler and listening devices used in Governmental surveillance, can be earmarked as an instrument of enlarging and imposing the Simulacra. The larger cyber world and the means of effective communication and representation are to be fathomed as the extensions of human sensory perceptions in the contemporaneous subset of make-believe.

---

[1]Matrix is a mainstream Movie made in Hollywood in the year 1998 which refers to a tug of war between the machines and the humans in a computer simulated world. Lana Wachowski, Lili Wachoswki, Release: 1999, United States.

[2]Jean Baudrillard, "Simulations," Massachusetts: MIT Press, 1983, Page no. 1–10.

Once again quoting from Jean Baudrillard, one can spawn a tedium of subjectivity which will soon make way for the nuances of objectivity and a partially empirical narrative. Still, Simulacra, also, posits the reality that a repetitive and monotonous transmission of hyperreality takes the steam out of the narrative and instead the lack of profundity of the make-believe, takes its place as a non-farsighted outcome of the imposition of a Big Data Universe. Jean Baudrillard can be quoted as, "Baudrillard's bewildering thesis, a bold extrapolation on Ferdinand de Saussure's general theory of general linguistics, was in fact a clinical vision of contemporary consumer societies where signs don't refer anymore to anything except themselves. They all are generated by the matrix. In effect Baudrillard's essay (it quickly became a must to read both in the art world and in academia) was upholding the only reality there was in a world that keeps hiding the fact that it has none. Simulacrum is its own pure simulacrum and the simulacrum is true."[3] The positioning of the Simulacrum is for itself, wherein the symbolism of the consumer sign boards comes much later, but the selfsame designation of the signboard reflects the all-pervading impact of the machine-generated make-believe.

It can be further cited as, "In his celebrated analysis of Disneyland, it is demonstrated that its childish imagery is neither true nor false, it is there to make us believe that the rest of America is real, when, in fact America is a Disneyland. It is of the order of the hyper-real and of simulation. Few people at the time realized that Baudrillard's simulacrum itself wasn't a thing, but a "deterrence machine," just like Disneyland, meant to reveal the fact that the real is no longer real and illusion no longer possible. But, the more impossible the illusion of reality becomes, the more impossible it is to separate truth from false and the real from its artificial resurrection, the more panic-stricken the production of the real is."[4] Thus, a "positivist skeptic" might be made to revel in the antagonistic notion that the effort at streamlining make-believe can render asunder the strength of a cunningly posited Simulacra. The non-functional nature of the Simulacra can be festooned upon the narrative here. The argument is that artificially created worlds and systems tend to end up as fanciful attempts at popularizing a reality which has to be utilized politically and systematically as the guiding narrative in a nation, commercial complex or a larger world. That makes for a great deal of political and national sense.

In an understanding of Marshall McLuhan, the Media theorist, *Medium is the Message*, seems to be the running dictat of the times. Marshall McLuhan considers the cell phone as the machine extension of an ear for human emancipation. The other notion being that a car is an apt extension of the four-wheeler, which can take a person across long distances in quick time with the least efforts. The spectacle is a personification of the eye which enhances human vision. Still, the idea that medium

---

[3]Ibid 1.
[4]Ibid 2.

is the message carries the real data content and can be dexterously managed and even contorted for various purposes.[5]

Marshall McLuhan writes, "Thus, with automation, for example, the new patterns of human association tend to eliminate jobs, it is true. That is the negative result. Positively, automation creates roles for people, which, is to say depth of involvement in their work and human association that our preceding mechanical technology had destroyed. Many people would be disposed to say that it was not the machine, but, what one did with the machine, that was its meaning or message."[6] Thus, Big data does in a way automate the mechanics of governing and rule making along with the running of industry. It is always the machine now in the contemporary context.

3 Comparative Idiom of Big Data in India and USA

One can hark back from the "life in the cables argument" to the reality of legislation and worktables machines. The Indian Big Data policy can be revisited with fitting comparisons with the American perspectives including the IT issues, policy, cyber security and trolling legislation if any one of them can be neatly deciphered. Big Data in India is destined to get bigger and better in 2017 as per analysis and reports. Three estimates in the Indian sphere are of great significance. The issue of sentiment analysis, customer engagement along with the idiom of bank management of accurate information that can constructively engage a deluge of data. Instances abound like, "Banks will seek the ability to access even more accurate data and in near real time, allowing institutions to be more cost efficient, and more profitable."[7] Also, increase in focus could also be observed in the following situations:

- There will be an increased focus on customer engagement and the customer experience versus traditional focus on customer service through the innovated application of big data and predictive analytics that allow organizations to provide better-customized services.[8]
- Institutions will seek the development of "sentiment analysis" technology that provides a gauge on individual customer satisfaction levels.[9] The research firm IDC reports that "Big Data is the shorthand label for the phenomenon, which embraces technology, decision-making and public policy. Supplying the technology is a fast-growing market, increasing at more than 30% a year and likely

---

[5]Marshall McLuhan, "Medium is the Message", URL: http://web.mit.edu/allanmc/www/mcluhan.mediummessage.pdf. (Online: Web) Accessed on 10 September, 2016.

[6]Ibid 1.

[7]Frank J Olhorst, "Big Data poised to Get Much Bigger in 2017," URL: https://gigaom.com/2016/12/01/xavient_2017/ (Online: Web), Accessed on 19 March, 2017.

[8]Ibid.

[9]Ibid 1.

to reach \$24 billion by 2016."[10] The firm further reports that "The McKinsey Global Institute, the research arm of the consulting firm, projects that the United States needs 140,000–190,000 more workers with "deep analytical" expertise and 1.5 million more data-literate managers, whether retrained or hired, by 2020."[11] Also, the practice of the Electoral and Parliamentary Presidential campaigns too rely heavily upon the idiom and the art of collation of Big Data which literally converts the electioneering processes of the parties into a war room. Department of Statistics in the state of Wisconsin informs about the knowledge collation exercise in the university campuses. For instance, the instrumentalities in the University of Wisconsin contends that "The campus knowledge-base is an improvement, where information is beginning to be organized. If this can utilize authentication so that individuals can drill down to more sensitive material as appropriate, then expansion of knowledge based website (kb.wisc.edu) to unit administrators would seem a wise investment."[12] This is an innovation in Big Data management and hiking the stakes of Big Data in US higher education system. It includes the conditionality's of huge backups, mammoth data storage along with design and multi-level curation.

In a manner, both structured and unstructured data take its toll on the theme of cyber security and cyber intelligence. India's cyber security too can be analyzed in the nom de plume of the larger issue of Big Data. The treacherous movement away from the data holder in the form of data theft is a novae element in the larger matrix of imbibing the role which a secure and stolid heft of Big Data can do to the present-day technology guzzlers of the day. The National Cyber Security Policy, also, defines cyberspace, "As an environment, where-in, there are interactions between people, software and services supported by worldwide distribution of Information and Communication Technology (ICT) of devices and networks."[13] The National Cyber Security Policy refers to the various incidents which can take place in the larger idiom of the transactional quid-pro-quo in the international system. The Indian policy document released by the Ministry of Communication and Information Technology contends that "A National and sectoral 24×7 mechanism has been envisaged to deal with cyber threats through National Critical Information Infrastructure Protection Centre (NCIIPC)."[14]

Computer emergency response team (CERT-In) has been designated to act as a nodal agency for coordination of crisis management efforts. CERT-In will also act as an umbrella organization for coordination actions and operationalization of sectoral CERTs. A mechanism is proposed to be evolved for obtaining strategic

---

[10]Steve Lohr, "Sizing Up Big Data, Broadening beyond the Internet," New York Times. June 19, 2013.

[11]Ibid.

[12]Department of Statistics, University of Wisconsin, "UW-Madison's relationship with Big Data".

[13]URL: http://meity.gov.in/sites/upload_files/dit/files/National%20Cyber%20Security%20Policy %20(1).pdf (Online: Web), Accessed on 19 February, 2017.

[14]Ibid.

information regarding threats to information and communication technology (ICT) infrastructure, creating scenarios of response, resolution and crisis management through effective predictive, prevention, response and recovery action.[15] The larger strategy initiated by the Government of India, also, involves the idea that there ought to be collaboration between the industry and the academia along with the establishment of Centers of Excellence in the realm of information and technology.

A timely commentary in the related context also indicates the sphere of national security and the freedom of privacy rights, which have been gaining traction since the last few years.[16] An, Institute of Defense Studies and Analysis (IDSA), release suggests about the indelible and all significant construct about the relationship between national security and privacy rights. The report suggests that "Indian Armed forces are in the process of establishing a cyber command as a part of strengthening the cyber security of defense network and installations. Creation of cyber command will entail a parallel hierarchical structure and being one of the most important stakeholders, it will be prudent to address the jurisdiction issues right at the beginning of policy implementation."[17]

The global debate on national security versus right to privacy and civil liberties is going on for long. Although one of the objectives of this policy aims at safeguarding privacy of citizen data, no specific strategy has been outlined to achieve this objective.[18] Thus, Big Data spawning, maintenance and sustenance signify the utilitarian forms which are crucial in the nature of the data function for the Defense sector in the nation. The idiom of intelligence collection, its analysis and distribution, leaves a lot to be desired but related scholarly and praxis articulations on the subject along with the idiom of national security, makes the Big Data paradigm to emerge as the significant idiom for data governance. Data too can be structured, non-structured and partially structured. The data present in the emails, mobile phone data and library and health records can be considered as privacy rights affected Big Data which can be compromised in the context of the national security concerns.

The American perspective is very clear about its preoccupations and privacy and other prescriptive sections. Steve Coffman writes in the Northern Illinois Law Review that "Section 702 of the Foreign Intelligence Surveillance Act (FISA) mandates warrantless electronic surveillance of suspected foreign communications by the National Security Agency (NSA). Section 215 of the 2001 USA Patriot Act, an amendment to FISA, also allows for a sweeping collection of domestic and foreign business records from private phone companies, to be queried with a

---

[15]URL: http://www.idsa.in/idsacomments/NationalCyberSecurityPolicy2013_stomar_260813 (Online: Web), Accessed on 19 March, 2016.

[16]Sanjeev Tomar, "National Cyber Security Policy 2013: An Assessment", URL: http://www.idsa. in/idsacomments/NationalCyberSecurityPolicy2013_stomar_260813 (Online: Web), Accessed on 19 March, 2016.

[17]Ibid.

[18]Ibid 1.

deferential showing of the information's relevance to foreign intelligence. The NSA may accordingly subpoena and buy data stored by private companies to conduct mechanical searches to find foreign intelligence. Meanwhile, companies like Facebook and Google sell analyzed information on individuals to advertisement companies, while research empires create massive data collection and analysis systems targeting consumers."[19] The American law is very clear and pithily laid out as far as the rights and functions of the Government duties are concerned. When it comes to the stringent upkeep of national security/homeland security aspirations and conditionality's, then, the Government agencies can impinge upon the legislations such as relevant sections in Foreign Intelligence Surveillance Acts, the much talked about Patriot and the Homeland Security Act. These are kinds of emergency scenario legislations, which are conditioned by the emergent threats to internal security and the rising threats of global terror apparatus including the funding of terror modules and transnational terror infrastructure.

National Security concerns have made it as a persistent poser before the scholars and practioners of Big Data alike. In the aftermath of the September 11 attacks, the international system and the American nation faced a national security catastrophe of mammoth proportions. It was made mandatory for the citizen surfers, audiences and readers to be mapped for their variously eclectic reading, surfing, phoning, traveling, and Cinema-going faculties. Every citizen with a South Asian and a Moslem origin was placed under an additional level of surveillance, wherein the personal faculties and engagements of the citizens were perilously begun to be pried upon. These American laws were soon recognized and universally castigated as being intrusive of privacy rights and personalized notions of right to liberty and happiness which lie embalmed in the grandiloquent American Declaration of Independence. The US Declaration of Independence proudly proclaims, thus, "We hold these truths to be self-evident, that, all men are created equal, that they are endowed by their creator with certain unalienable rights, that among these are Life, Liberty and the pursuit of Happiness."[20] Thus, freedom lies enshrined in the dialogue of the American Dream and its Founding fathers. They ate the trailblazing icons as the curators of the ideals of liberty, equality and egalitarianism.

Still, it might come as a surprise that the US Constitution does not constitute a well-prescribed right to privacy. The fragment of law is hinted upon, and there is no explicit mention of the entire thematic percept. Still, scholars and American Founding fathers such as James Madison have always maintained likewise in the direction of the right of privacy issues. They have contended that personages who are citizens of USA should be free from sustaining their home and hearth and

---

[19]"Modern Private Data Collection and National Security Agency Surveillance: A Comprehensive Package of Solutions Addressing Domestic Surveillance Concerns" Northern Illinois University Law Review, Vol. 34, 2014.
[20]In the Constitution of the United States, Article II (sec. 1, cl.3).

cannot be forced into housing soldiers as it was the practice during the wars and the legendary Civil war.[21]

The Fourth Amendment rights have always existed in USA and the attendant jurisprudence has always been present. One can delve inside the idiom of a land-mark wire-tapping case in the US Supreme Court which adjudicated upon the Olmstead versus USA in June 4, 1982. The Justice Brandies' judgment declared with aplomb about the individual personages to be "let alone" specifically in the context of wire tapping. The case details can be surmised as follows, "The makers of our Constitution undertook to secure conditions favorable to the pursuit of happiness. ... They conferred, as against the government, the right to be let alone— the most comprehensive of rights and the right most valued by civilized men. To protect that right, every unjustifiable intrusion by the government upon the privacy of the individual, whatever the means employed, must be deemed a violation of the Fourth Amendment."[22] Personages could travel to any part of the nation and also were supposed to be free from unreasonable personal searches and had the right to own personal possessions. National security and national interest make an iota of compromise necessary for the upkeep and sustenance of the homeland concerns.

In an excerpt from the historical novel, written by the renowned American historical playwright, E. L. Doctrow, one can decipher the lengthening shadows of the trenchant civil war and the fear of the people in the Southern city of Georgia which faces unwanted possession by the Northern Unionist forces of General Sherman. The excerpt of the novel highlights the initial cast upon the right to privacy and possessions and the American people. "And, as they watched, the brown cloud took on a reddish cast. It moved forward, thin as a hatchet blade in front and then widening like the furrow from the plow.... When the sound of this cloud reached them, it was like nothing they had ever heard in their lives. It was not fearsomely heaven-made, like thunder or lightning or howling wind, but something felt through their feet, a resonance, as if the earth was humming. Then, carried on a gust of wind, the sound became for moments a rhythmic tromp that relieved them as the human reason for the great cloud of dust".[23] In the same sense, the tranquil existence and the secure shibboleths of man and citizens in USA and other parts of the Globe are threatened by the invalid possession and insidious surveillance and possession of Big Data, which is the new "tangible", apart from the physical possessions of houses, gardens, flats, porticos and bank balances owned by indi-vidual personages. In emergency scenarios, the law does not refer too much.

---

[21]E. L. Doctrow, "The March," New York: Random House, 2006.

[22]US Supreme Court, Olmstead versus United States, 277 US 438 (1928).

[23]Ibid.

4   The Trolling on Social Media: When the Real-Time Meets the Make-Believe

The word, trolls, does nor refer to here, with the advent of the cartoon characters which are very popular among children. Trolls are an insidious expression of angst and frustration by antagonist view and perception holders in the cyber world. In several nations, such as UK, there is a conviction rate of trolls committed on the social media. In UK, five trolls are convicted every day. The question which can be delved inside is the idea that can the entire debate be termed as trolls being criminal offenses. The conviction rate for the trolls has been augmented around ten folds since the times, indecent, vulgar and obscene messages began to be disbursed and adjudged as being inimical to the Internet users. A report of the newspaper, Telegraph, can be cited which informs about the statistical nom de plume of trolling on social media which may be offensive to some Internet users. This is another facet of the reconfiguration and the maintenance of the Big Data spectacle.

The issue of Internet trolling can be surmised as, "It is a crime under the Communications Act to send 'by means of a public electronic communications network'" a message or other material that is "grossly offensive or of an indecent, obscene or menacing character. Previously little-used, Section 127 has come to prominence in recent years following a string of high-profile cases of so-called trolling on social media sites. It can also cover phone calls and emails, and cases of "persistent misuse" that causes the victim annoyance, inconvenience or needless anxiety."[24] Thus, any such social media vitiation which causes vitiation of the entire socio-psycho make up of an Internet user in an offensive and forceful manner, can, be booked under the law but in the case of India, such guidelines are not clearly laid down.

The same Telegraph report goes ahead with referring to the increasing trolls in the Internet world and also the augmented rate of conviction of the trollers and social media abusers. The report contends that "Professor Lilian Edwards, Director of the Centre for Internet Law and Policy at the University of Strathclyde, said the rise in prosecutions under Section 127 reflected the surge in use of social media. This was a relatively obscure provision before the internet. You would have been talking about poison telephone calls and there were relatively few of those. It is obviously related to what has happened with social media."[25]

Earlier on people would attempt to rile through "poison phone calls" and other, "call threats," but with the advent of the idiom of immediacy in the realm of social media, the various media lords in an individual capacity have been empowered in the light of the Collectivization and Mobilizational power of the social media and the attendant activists. This is definitely a problem related to the Big Media and Big Data wherein privacy rights and other issues can be managed by the State.

---

[24]Agency, "Five Internet Trolls a Day Convicted in UK as Figures show ten fold increase," Telegraph, 24 May, 2015.
[25]Ibid.

The women editor of the newspaper, the Telegraph, Emma Barnett writes that "Online actions have real life consequences. But we have tended to ignore that hard truth. Collectively we've told people who have been the victim of trolling or other injustices online to just get a grip and stay schtum. After all, it's only happening on the internet. Just turn the computer off and walk away. Right?"[26] In the make-believe protection and the all-pervading aura of the Big Data, we tend to forget the idiom that Big Data pyrotechnics have to come down to the real-time world also, only then the freedom which one enjoys and misuses on the Facebook, Twitter, and Instagram posts can be equated to a real-time happening or a real-time crime, which is a bookable offense.

5   Some Inferences

The inferences which can be pithily drawn amount to the following. As an instance, there is a need to decipher the Simulacra-like impact of the Web world and the cyber concerns. The ideational aspect of the make-believe world needs to be tempered with the cyber regulatory aspects of the Indian and the American situation in light of their anti-trolling legislations, with regard to their comparisons. In India, the privacy laws are not in a mature state of affairs, while the American situation already allows for a regulatory role by the Institutions such as the judiciary and the executive in milieu of the cyber transactions and the general sentiment of the cyber context. Also, the actual rate of conviction in the world of social media needs to be delved inside. One of the inferences happens to be that the right to privacy and the idea of liberty and happiness of the commoners needs to be underlined against the interventions by the State and the crime mongers in the cyber world.

One typical aspersion can be on the national security concerns of the nations such as India and USA. Some of the individual personages masquerade as the victims of the establishment and attempt to spread a clarion call in the name of individualism, anti-national and anti-societal dissent. The balance need not be reached between the sections clamoring for more liberty and more of a checks and balances system for ensuring the safety, security, and stolidity of the nation. Being functional and efficient are the idiom of the day which emphasizes the idea of bringing about a normalcy of realism in the narrative buttressed by a legal and constitutional perspective of the laws relating to the cyber world and the social media.

---

[26]Emma Barnett, Madeline Mc Cain "Twitter Troll Death: Trolling is Never a Victimless Crime", URL: http://www.telegraph.co.uk/women/womens-life/11144435/Madeleine-McCann-Twitter-troll-Brenda-Leyland-death-Trolling-is-never-a-victimless-crime.html (Online: Web), Accessed on 12 September, 2016.

6   Conclusion

Information is always beautiful. Harking back to the American context, the National Security Agency informs that it had tied up with a renowned data security firm called as "Sqrrl". This goes a long way in enabling nerd-like analytics agents in a matrix-ordained cyber sphere in order to stay as the staple vanguard of the American security establishment. Some studies portray that how correlations and Googled worlds are taken up in order to secure the Web security systems in USA and how the algorithm-enabled Knowledge bases strengthen the dictum of Knowledge is Power.[27] "The approach Sqrrl takes in organizing and processing a year's worth of all kinds of machine, and network log data is based on visualizations of the links–or associations–between elements. Google is probably the purveyor of the most popular graph algorithm in the world, which is, of course, called Page Rank. A lot of the way that Google does its searches on its semi-structured data is using graph algorithms that look at the links between Web pages to determine the significance of Web pages." Thus, here too the core argument of the paper works that the real-time happenings in the real security sphere gels with and gets equated with the blogosphere and the Simulacra. Thus, what must catch the eye of the storm is the notion that "Make-believe," meets the matrix germ of the cyber world. Finally at the end of the day, it is the notion of functional ease and convenience which labors, under the idiom of the finesse and the sophistication of the Big Data theme, in India and systems abroad.

---

[27]Robert Richardson, "NSA's big Data Security Analytics Reaches the Enterprise With Sqrrl". URL: http://searchsecurity.techtarget.com/opinion/NSAs-big-data-security-analytics-reaches-the-enterprise-with-Sqrrl (Online: Web), Accessed on 1 September, 2016.

Evolving an Industrial Digital Ecosystem: A Transformative Case of Leather Industry

Latha Anantharaman and M. R. Sridharan

1 Introduction

Leather manufacture, considered as an art, is a process known to man since time immemorial. The processing underwent several changes based on scientific investigations over the period and at present is considered a modern technology. In the early stages of leather processing, only natural products were used for making leather. However, today a variety of chemicals and synthetic auxiliaries are employed. Leather processing, in general, utilizes copious amount of water. Many organic and inorganic chemicals such as acids, alkalis, tanning and retanning agents, fat liquors, dyes, polymeric binders, plasticizers, feel modifiers, top coats, and organic solvents are used in the wet processing.

The basic raw materials for this industry, (viz., hides/skins), are by-products of the meat industry. The hides and skins are dispatched and stored under wet or dry salted conditions. The conversion of putrescible hides and skins into stable leather involves a variety of complex chemical and mechanical operations based on sound scientific principles. Though the chemicals and the operations are not under high pressure or temperature or subject to hazardous reactions, it can be said that unsafe storage, handling, and operations may lead to accidents.

Leather industry occupies a place of prominence in the Indian economy in view of its massive potential for employment, growth, and exports [1]. The Indian leather sector meets 10% of global finished leather requirement. There has been an increasing emphasis on its planned development, aimed at optimum utilization of available raw materials for maximizing the returns, particularly from exports. The

L. Anantharaman (✉)
Knowledge Resource Centre, Central Leather Research Institute, Chennai, India
e-mail: platha@clri.res.in; lathaclri@gmail.com

M. R. Sridharan
Department of Chemical Engineering, Central Leather Research Institute, Chennai, India
e-mail: mrsridharan@clri.res.in; mrs_clri@yahoo.co.in

© Springer Nature Singapore Pte Ltd. 2018
U. M. Munshi and N. Verma (eds.), *Data Science Landscape*,
Studies in Big Data 38, https://doi.org/10.1007/978-981-10-7515-5_18

Fig. 1 Percentage of shares of global countries in Indian leather/leather product exports (2014–15). Source of the data www.leatherindia.org

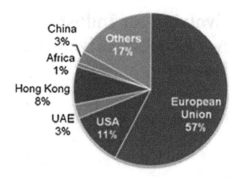

Indian leather industry employs about three million workforces. The leather industrial sector comprises of: (a) tanneries (where raw hides and skins are converted into leather) (b) factories transforming leather into a variety of consumer products such as footwear, garments, and assorted leather goods such as wallets, passport cases, key chains, handbags, and brief cases [2]. The leather industry, numbering around 2500 units, is mainly concentrated in three states of India, viz. Tamil Nadu, West Bengal, and Uttar Pradesh. Nevertheless, most of the major units (approximately 60%) are situated in the state of Tamil Nadu, in the southern part of India. The main production centres for leather and leather products manufacture in India are the following:

1. Tamil Nadu—Chennai, Ambur, Ranipet, Vaniyambadi, Trichy, Dindigul, and Puducherry.
2. West Bengal—Kolkata.
3. Uttar Pradesh—Kanpur, Unnao, Banther, Agra, and Noida.
4. Punjab—Jallandhar.

Tamil Nadu accounts for about 40% of India's exports and has about 60% of tanning capacity [2]. A typical tannery employs both unskilled and skilled workers as well as qualified supervisors and technologists. India has an abundance of raw materials with access to 20% of world's cattle and buffalo and 11% of the world's goat and sheep population [3]. India's leather industry is set to grow exponentially over the next five years with a growth target of 50% in exports from 2016 to 20 [4]. The percentage of shares of global countries in Indian leather/leather product export is depicted in Fig. 1.

2 Sustainability of the Industry

The future growth of Indian leather sector has to be technology driven. In order to meet the diverse technological needs of the Indian leather sector manufacturing units, a well-knit network of private R&D and public funded research institutions is essential. A good blend of basic and applied R&D efforts is necessary to achieve the short and long range of technological goals as envisaged for the Indian leather sector beyond 2000. The technology, design, and R&D inputs to the leather and applied product sectors will have to be essentially driven by the several factors [5]:

- Concern for environmental and occupational health aspects, minimization of material and energy wastages, and usage of indigenous resources.
- Highest quality coupled with consistency.
- Water conservation through better management and recycle practice.
- Automation to enhance the capabilities and productivity of the human skills.

Many chemicals are transported and used for leather processing. Chemical Inputs into leather sector have to be thoroughly assessed for Life Cycle Analysis (LCA). LCA [6] is a tool that can be used to evaluate the potential environmental impacts of a product, material, process, or an activity. An LCA is a comprehensive method for assessing a range of environmental impacts across the full life cycle of a product system, from materials acquisition to manufacturing, use, and final disposition. It is widely recognized that at all levels of development, small- and medium-sized enterprises (SMEs), have a significant role to play in economic development in general and in industrial development in particular. SMEs form the backbone of the private sector, make up over 90% of enterprises in the world and account for 50–60% of employment [7].

To set up an effective safety management approaches, SMEs need to be involved in initiatives aimed at fostering chemical safety management and chemical risk information throughout the value chain. These initiatives need to engage a wide range of stakeholders including workers, transporters, distributors, customers, authorities, nearby communities, and other businesses including larger companies [8]. Creation of eManuals and digital supply chains will help in realizing this situation. Green industry promotes sustainable patterns of production and consumption, i.e. patterns that are resource and energy efficient, low carbon and low waste, non-polluting and safe, and which yield products that are responsibly managed throughout their life cycle.

3 Proposed Model for Leather Industry as Industry 4.0 (I4.0)

Industry 4.0 is being driven by:

- Digitization and integration of vertical and horizontal value chains.
- Digitization of product and service offerings.
- Development of new digital business models and customer access platforms.

The nine pillars of I4.0 (Fig. 2) as generally agreed upon are [9]

Fig. 2 Nine pillars of industry 4.0 (I4.0)

Flow Chart 1 Flow chart for
establishing industry 4.0

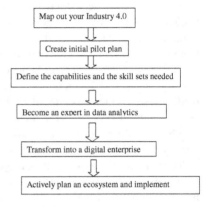

"Digitalization" is defined as the process of moving to a digital business that will lead to new revenue opportunities and change a business model. The various steps for establishing industry 4.0 are given in the following Flow Chart 1.

Various components that are required for establishing industry 4.0 (I4.0) for a conventional leather industry to transform into an digital eco-model [10] are listed below:

- e-Tannery: usage of computers, intranet, Internet, videoconferencing for linking tanneries globally and projecting tanneries in digital space.
- Online information systems on raw material availability, inventory on chemicals and machinery, processes, products, and emission.
- New digital business models such as e-Commerce, m-Commerce, new standards and norms, marketing issues, promotion of network practices.
- Enterprise resource planning (ERP) for strategic and operational planning, manufacturing, management of materials, quality management, finance, logistics management, maintenance management and sales and distribution.
- e-Governance, a model for digitalized management of the leather industry.
- Digital library information system on leather products, "Who and Where" information linking with national and global data.
- Expert systems for forecasting.

- e-Learning—multimedia-based online training techniques and centres for building up knowledge society and human resource development and novel mode of education for development of twenty-first century skill sets such as eSTEM.
- Robotics and process control techniques for automation of tannery processes and product manufacture.
- Application of geographical information system (GIS) in appropriate setting up of an eco-friendly tannery (that will lead to least stress on environmental issues).

3.1 Modernization Using New Technologies of Information and Communication [11]

List of digital technologies that contribute towards industry 4.0 is given in Table 1. The following sections briefly describe the methods for implementing I4.0.

3.2 Geographic Information System (GIS)—e-Governance, a Model for Digitalized Management of Leather Industry

A geographic information system is a computer system designed to capture, store, manipulate, analyse, manage, and present a variety of spatial and geographical data.

Table 1 Industry 4.0 framework and contributing digital technologies

Software	Hardware	IT
• New organizational concepts • Modern process techniques • Low waste technologies • Simulation and modelling • Image processing • Artificial intelligence • Neural network • Expert systems • Pattern recognition • Robotics • Process recipes • Business process reengineering	• Plant and machinery upgradation • Instrumentation and controls • Temperature, pressure, pH controls • Inprocess systems for BOD, COD, TDS control • Smart sensors • Embedded systems for specific purposes like leather defects analysis, calibration systems, etc.,	• e-mail • Internet and intranet applications • Web services • Videoconferencing • OLTP • ERP • Data warehousing • Data mining-Big Data analytics • e-Commerce, m-Commerce • e-Learning • Workflow • e-Procurement • e-Tannery • Virtual laboratory • Data centre • Cloud computing

Table 2 Tannery details of Tamil Nadu data in Excel format

Tannery Details in Tamil Nadu			
Total Tanneries : 790			
Name	City, Cluster & State	Items	Stage
A & A Leathers No.437/2B, C. Abdul Hakeem Road, Melvisharam, Ranipet. Pin- 632 509 Tel. Ph- 04172 - 267317	City- Melvisharam Cluster- Ranipet State- Tamil Nadu	Sheep Skins Goat Skins Cow Calf Skins Buffalo Calf Skins	Raw to Semi-Finish
A Abdul Hai Tanning Company S F No. 400/1A3,V Kota Road, Pernambut. Pin- 635 810 Tel. Ph- 04171 - 232606	City- Pernumbut Cluster- Pernambut State- Tamil Nadu	Buffalo Hides	Raw to Finish
A M Leathers No.103, TM Road, Nagalkeni, Chromepet, Chennai. Pin- 600 044 Mob- 9789076365 / 9042305386	City- Chennai Cluster- Pallavaram State- Tamil Nadu	Sheep Skins Goat Skins Cattle Hides Cow Calf Skins	Semi-Finish to Finish
A One Leather No.1056/B, Konamedu, Vaniyambadi. Pin- 635 751 Tel. Ph- 04174 - 232138 Fax- 04174 - 232527 Mob- 9790021588 / 84 Email- a.oneletVNB@yahoo.com	City- vaniyambadi Cluster- Vaniyambadi State- Tamil Nadu	Sheep Skins	Raw to Finish
A P J Associates No.168, SIDCO Industrial Estate, SIPCOT, Ranipet. Pin- 632 403 Tel. Ph- 04172 - 247243	City- Sipcot Ranipet Cluster- Sipcot Ranipet State- Tamil Nadu	Sheep Skins Goat Skins Cattle Hides	Semi-Finish to Finish
A P K Chemicals No.44A, MGR Road, Nagalkeni, Chromepet, Chennai. Pin- 600 044 Tel. Ph- 044 - 22385083	City- Chennai Cluster- Pallavaram State- Tamil Nadu	Sheep Skins Goat Skins Cattle Hides Cow Calf Skins	Semi-Finish to Finish

Source On_going survey and creation of database on tannery details of Tamil Nadu by Economics Department, Central Leather Research Institute, Adyar, India

GIS technology can be used for scientific investigations, resource management, and development. The benefits of using a GIS are numerous, because GIS takes into consideration many different factors to help build an efficient and organized way of presentation of spatial data. Table 2 gives the details of tannery in Tamil Nadu district. It provides various information such as the types of jobs carried out, the stages of leather processing, e-mail, telephone number, and the addresses of the tannery.

A GIS displays the markers based on two important parameters such as latitude and longitude. These two attributes have to be added to the excel sheet. Latitude and longitude can be obtained from the address of a place by a process called Geocoding. The Google helps in converting the addresses into their respective latitude and longitude using, Geocoding API.

A map is displayed with markers which denote the respective tannery's location. Upon clicking the marker, an infobox is opened displaying the details such as name, address, owner of the tannery, exports, imports, etc. Various other graphs, pie-charts, and bar graphs can also be created to show the number of tanneries in each state, growth in volume of exports/imports, etc. The result of building a GIS would lead to a Web portal that can display information about all the tanneries present in India as well as elsewhere in the world.

The scope of this data centric application is immense, as it can be extended to tanneries for any number of countries in the world, or the whole world itself, depending on the available data and requirements. Further, it can be used not only for tanneries but also as a GIS for any large organization such as a supply chain wishing to know its branches in various cities, etc.

A GIS created for tanneries in India would be of paramount importance to the leather industry and to the government. It ensures that the data they have so far obtained are not lost by preserving it in a digital format A number of statistics can also be displayed using infographics in form of charts, graphs, etc., which are helpful to determine the concentration of the leather industry in a specific region and how to improve them.

The steps involved in creating a GIS for tanneries in India are as follows:

1. Extraction of the data from HTML code.
2. Digitization of the available data from books/other sources.
3. Creating and populating the database using MySQL.
4. Outputting the data as XML using PHP.
5. Creation of the map.
6. Creation of custom markers to display the locations of tanneries.

A sample of creating GIS for tanneries based on the above steps is reflected in Fig. 3.

For the leather industry in general, and the owners of the tanneries, researchers, and scientists of leather in specific, querying the details of various tanneries across India for their requisite purposes is crucial. Since there are numerous tanneries present in India, the method followed until now involved taking a survey throughout the country and publishing the information in the form of books.

In order to promote greener and cleaner technology, details about the chemical oxygen demand (COD), biochemical oxygen demand (BOD), total dissolved solids

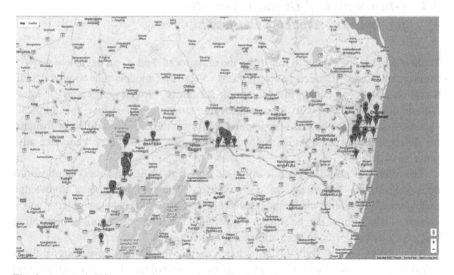

Fig. 3 A typical GIS representing tanneries in Tamil Nadu

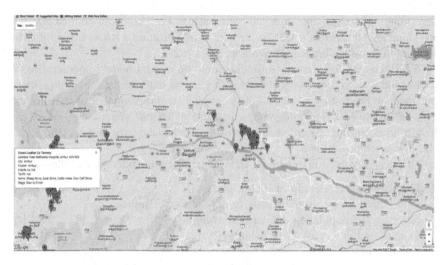

Fig. 4 Details about a tannery in Ambur city using GIS tool

(TDS) from each tannery can be monitored using this GIS tool (Fig. 4). For example, using GIS, a single map could include sites that produce pollution, such as factories, and sites that are sensitive to pollution, such as wetlands and rivers. Such a map would help people determine where water supplies are most at risk.

With the intention of promoting I4.0 model, it is important to capacitate human resources using digital ecosystem. The same is being detailed below.

3.3 e-Learning and Digital Eco-model for Industry-Academy Partnership

Technology has the power to transform the class room, and it can pave new ways of training and ideas. Blended learning refers to learning models that combine traditional classroom practice with e-Learning solutions. Providing libraries of information that can be used to train employees, students will reduce the duplicate expenses being spent because the learning objects can be used in different formats, for different purposes. With distributed technologies, learning can take place anytime, anywhere.

The Web can now integrate learning and mission-critical business applications delivering timely knowledge to each desktop. The end result is a knowledge management structure which includes an inventory of knowledge objects and a system in which these can be shared. Knowledge has long been a tool of innovation and a key driver for economic and social development.

The "21st Century Skills" [12] includes digital age literacy consisting of functional literacy, visual literacy, scientific literacy, technological literacy, information literacy, cultural literacy, and global awareness, inventive thinking, higher-order thinking and sound reasoning, effective communication, and high productivity. Various competencies must be developed throughout the educational system for ICT integration to be successful.

For concerted efforts in capacitating human resources, there is a need for content development using standardized procedures.

3.4 e-Learning: Content Development and Standards

e-Learning delivers instruction anytime, at any place and in any combination desired by the learner. The salient features of e-Learning are as follows: anytime, anywhere availability, efficiency, and flexibility and inexpensive worldwide distribution. e-Learning is best defined as 'content' with 'standards' that addresses the teaching objectives; 'tools' to develop, manage, and deliver the learning; and 'technology' infrastructure that will support the tools and facilitate the delivery of content. SCORM [13] is defined as Sharable Content Object Reference Model. SCORM is a set of technical standards for e-Learning software products. A learning management system (LMS) provides the platform for the enterprise's online learning environment by enabling the management, delivery, and tracking of blended learning (i.e. online and traditional classroom). Adobe presenter, Articulate, Lectora, etc., are some of very popular SCORM complaint content development software which empowers the subject matter experts (SME) to create content with little programming knowledge.

A pilot scale e-Learning system [14].

The pilot scale e-Learning system (digital ecosystem) has the following features:

- Integrate traditional lecture-based teaching with online learning materials and communication technologies through videoconferencing, wherever essential.
- Enhance teaching effectiveness using rich multimedia.
- Facilitate e-Learning content inside the campus and the study centre.
- Enable subject matter experts to rapidly create e-Learning content, in—house, using simple technologies and tools saving time, money, and effort.
- Provides anytime, anywhere learning for large audiences, spanning rural and global students, and working professionals.
- Utilize resource personnel from industries and higher learning institutions.
- Provide guided practice sessions through self-learning with an intensive online training programme.
- Provide eBooks, digital library, virtual labs, video-based curriculum, video-conferencing and multi-language support.
- Create a knowledge centre with multiple knowledge channels, with a long-term vision of creating a knowledge hub, connecting to national and global knowledge network.

4 Major Steps in Creating the Infrastructure

Based on the prototype developed, the following steps are involved in creating the infrastructure.

- Phase 1: Establish an e-Learning environment.
- Phase 2: Content development.
- Phase 3: LMS selection and implementation.
- Phase 4: Upload the content.
- Phase 5: Tracking, document management.

5 Set-Up of the Data Centre Running Distance Learning Programme

With these concepts, one can provide distance learning programme across the continents and one can also provide the learning material in multiple languages. Figure 5 describes a data centre set-up running a distance learning programme. With the advent of cloud computing, this can also be given as a software as a service (SaaS) model to provide a cost-effective solution.

The following pictures illustrate how student can login to the learning management system (LMS) on an Intranet by using appropriate password for accessing their course material (Fig. 6).

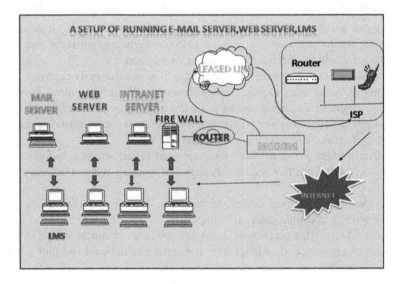

Fig. 5 Schematic representation of operational data centre

Fig. 6 View of student authentication system for accessing the course material

A fool-proof system of e-Governance decision support system is required to make the entire digital ecosystem for leather industry a robust one, and the same is detailed below.

6 e-Governance-Sustainable Development-Environmental Science, Technology, Engineering, and Mathematics (eGOV_SD_eSTEM)

Electronic governance (eGov) research studies the use of information and communication technologies to improve governance processes. Sustainable development (SD) research studies possible development routes that satisfy the needs of the present generation without compromising the ability of the future generations to meet their own needs. Developing an e-Governance system on environmental issues will help an organization to increase its efficiency in natural resource management and diminish wastes and emissions.

A green economy is likely to depend crucially on innovation, in particular eco-innovation. Global carbon trading will be a cost-effective tool to significantly cut greenhouse gases. Carbon credits are a key component of national and international emission trading schemes. They provide a way to reduce greenhouse emissions on an industrial scale by capping total annual emissions and letting the market assign a monetary value to any shortfall through trading.

Using data, the analysis is carried out with the following equations for finding the relationship between gross domestic product, carbon credit, and energy intensity.

Table 3 Technological options for cleaner leather processing

Process	Improvements implemented	Process	Improvements implemented
Pre-soaking	Mechanical desalting Improvised manual desalting	Deliming	Organic acid-based deliming, carbon dioxide based deliming Recycling of delime liquor
Soaking	Use of enzymes in soaking Countercurrent soaking	Pickling	Recycling of pickle liquor Salt free organic acid pickling system
Liming and unhairing	Recycling of lime liquors Enzyme assisted unhairing Hair saving technique	Chrome tanning	Closed pickle-tan loop Chrome recovery and reuse High exhaust chrome tanning

Source www.clri.org

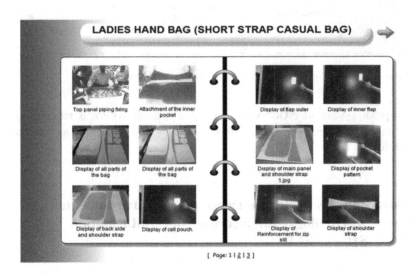

Fig. 7 Skill-based training programme: a snapshot

$$\text{Carbon Intensity} = \text{Energy Intensity} \times \text{Fuel Mix}$$

Carbon dioxide (CO_2) emissions are often hypothesized to follow environmental Kuznets curve model: $CO_{2t} = \alpha + \delta t + \beta 1 GDP_t + \beta 2 GDP_{2t} + \varepsilon t$ where CO_{2t} is CO_2 emission per capita at time t, GDP_t is GDP per capita at time t, $t = 1, 2, \ldots, T$, t is the time variable, and GDP represents the gross domestic product. STEM represents the fields of study in the categories of Science, Technology, Engineering, and Mathematics. eSTEM represents environmental STEM. Technological options for cleaner leather processing are listed in Table 3. A SCORM complaint digital training course for skill-based training programme on leather bag making that was developed using Lectora package is illustrated in Fig. 7.

A model of the levels of environmental performances is shown in Fig. 8. The first axis is the time axis, the product's lifetime with its phases in planning,

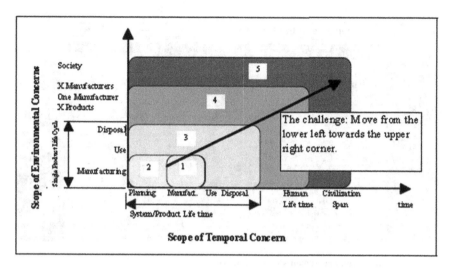

Fig. 8 Levels of environmental performance. *Source* Cleaner production and industrial ecology: dynamic aspects of the introduction and dissemination of new concepts in industrial practice by By Leenard Willem Baas, Leo Baas (2005, p. 89), Eburon Academic Publishers

manufacturing, use and disposal, human lifetime, and the civilization span. The second axis indicates the scope of the environmental concern, ranging from a single product life cycle, to x products within one manufacturer and towards x manufacturers and the society.

Design of eGov_SD_eSTEM portal is presented in Fig. 9. eSTEM education can be implemented by the method that was described in detail under Sect. 3.3.

We developed a carbon footprint calculator, and it calculates the carbon footprint of the various steps involved in the tanning process. Tables 4 and 5 are used for the carbon footprint calculation. The calculator uses HTML forms to input the data from the user and displays the result in a separate HTML page. The calculations are done using PHP [15] by taking the data entered using POST [16] method, and then, the required calculations are performed, and the net carbon footprint is printed out.

List of technologies for in-plant pollution reduction and end-of-pipe treatment for leather sector [19] as developed by Central Leather Research Institute (CLRI) is listed below:

- Chrome recovery and reuse.
- UASB technology complete with sulphur recovery plant.
- Biomethanation for solid waste disposal.
- Chemo autotrophic activated carbon (CAACO) system for wastewater treatment.
- Removal of total dissolved solids (TDSs) in tannery effluents.
- Waterless chrome tanning process.
- Zero wastewater discharge technology through electro-oxidation.
- Technology for wastewater treatment.

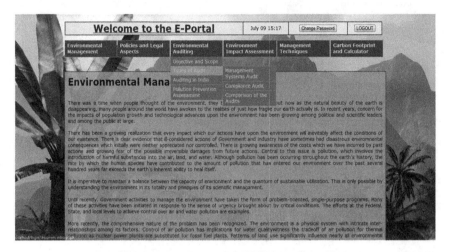

Fig. 9 System design and screen shot: a prototype

Table 4 Sample values for CF calculation due to tanning processes [17]

Process step	Amount of water (m³/tonne)	pH –	COD kg COD/m³	N_{kj} kg N/m³	Cr kg Cr/m³
Pre-tanning					
Soaking	4–6	6–9	30–40	1–1.5	–
Unhairing, liming	5–9	12–13	40–60	3–5	–
Fleshing	1–3	–	–	–	–
Deliming, bating	5–7	8.5–9	5–8	3.5–4	–
Tanning					
Chrome tanning	0.5–1	3.8–4	2–3	0.3–0.6	0.5–5
Pressing	0.4–0.6	3.6–4.5	1.2–1.8	0.11–0.22	0.5–5
Neutralization	1–1.5	4.5–4.7	2.5–3	0.5–0.8	0–1.0
Dyeing, fatliquor	3–4.5	3.8–4.5	5–6	0.2–0.3	0–5.0
Finishing					
Drying	3–6				
Finishing	1–2				
Cleaning	5				

Green management system comprises of environmental management systems (ISO 14001 certified), green waste reduction techniques comprising of process, product redesign, reduce, remanufacturing, that leads to good business results in terms of costs, lead times, quality, market position, reputation. Green supply chain management (GSCM) improves supply chain visibility and smaller carbon footprint and saves cost. Information technology can optimize transportation planning routes and ensures that goods/services are delivered in the most energy efficient and

Table 5 NCV and CO_2
emission factors of different
types of fuel used for
estimation. Look up table for
CF calculation for transport
related to logistics [18]

	NCV (Tj/kt)	CO_2 EF (t/Tj)
Coking coal	24.18	93.61
Non-coking coal	19.63	95.81
Lignite	9.69	106.15
Diesel	43	74.1
Petrol	44.3	69.3
Kerosene	43.8	71.9
Fuel oil	40.4	77.4
Light distillates	43.0	74.1
CNG	48	56.1
LPG	47.3	63.1
Lubricants	40.2	73.3

Note: *NCV* net calorific value; *EF* emission factor; Tj $= 10^{12}$ J
1 J $= 2.39 \times 10^{-4}$ kcal

cost-effective manner. Automation of the transportation planning process [20] using GIS (Fig. 10) enables transportation managers to mitigate the effects of unexpected events. Information technology can help in streamlining business processes and thereby enable reduction in resources usage while executing business processes.

Based on these experiments, the logistics data generated are used for digitization and integration of vertical and horizontal value chains for establishing GSTM.

6.1 Enterprise Resource Management

An Environmentally Responsible Business Strategy (ERBS) with mobile technologies can help organizations achieve socially responsible goals of reducing green house emissions, reducing physical movement of men and materials, and recycling of materials and many more. An integrated ICT model using cloud computing for networking of tanneries and academic institutes is proposed for SME's and MSME's (micro, small, and medium enterprises).

ICT enables a framework of new business model such as lean, green manufacturing, and zero waste technologies for the tanning sector. Green and lean manufacturing systems help to minimize environmental impact of manufacturing processes and products. An integrated lean and green manufacturing system model, zero waste manufacturing are proposed as a solution for economically and environmentally sustainable manufacturing. The outcome of lean model results in reduced inventory levels, decreased material usage, optimized equipment, reduced need for factory facilities, increased production velocity, enhanced production flexibility, eliminating waste throughout the production process, employee involvement, supply chain investment, and so forth.

Pollachi to Kerala - Transportation route of the animals

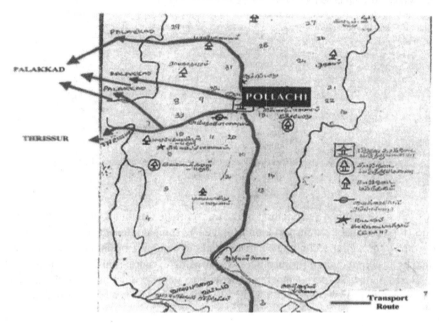

Fig. 10 Transport route of animals [21]

Supply chain operations reference (SCOR) model helps to identify potential areas for green initiatives in the supply chain and align green initiatives with the strategic objectives of the company. Industrial ecology attempts to look at industrial systems as ecosystems, whereby the waste of one process becomes the raw material of another. Automobile leasing, chemical leasing, use of environmental check list in product design, use of green index rating system on materials, and processes for product design and purchasing which lead towards low carbon footprint economy are facilitated by Green ICT. Schematic diagram of zero liquid discharge process is displayed in Fig. 11. Supply chain in transporting animals to slaughter house is given in Fig. 12.

7 A New Business Model at CLRI for Digital Datasets

In the backdrop of above discussions, CLRI produces lot of digital datasets in various forms and formats that could be part of so-called Big Data regime. These datasets are structured, semi-structured, or un-structured. Big Data represents datasets whose sizes vary from terabytes to zettabytes, and it has one or more of the

- Water recycle and reuse method based on zero wastewater discharge principle for pre-tanning operations standardized

- Water consumption levels reduced from 17 to 1.7 L/kg of hide (raw to wet blue)

- Technology demonstrations in progress

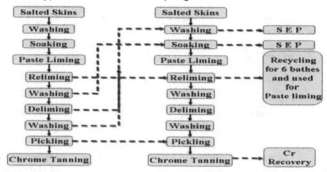

Fig. 11 Zero liquid discharge process steps [22]. *Source* www.clri.org

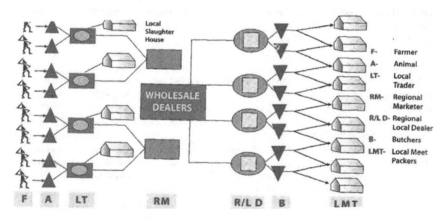

Fig. 12 Supply chain in transporting animals to slaughter house [21]

following characteristics—high volume, high velocity, and high variety. The datasets (Table 6) generated are maintained by the research institute, using advanced analytics techniques such as text analytics, machine learning, predictive analytics, data mining, statistics, and natural language processing. Businesses can analyse previously untapped data sources independent or together with their existing enterprise data to gain new insights resulting in significantly better and faster decisions.

Table 6 Data sources for Big Data administration

Raw material availability	Global live stock data	Global production of hides and skins	Global tanning and product sectors	Global trade in leather sector	
Econometrics					
Chemical inputs					
Inventory reach, Hazop	Chemical processes classified according to tannery process	Chemical classification	List of chemical industries	Safety aspects	
Physical inputs					
OR, cellular manufacturing	Machinery	Processing machineries	Finishing machineries	Manpower	
Processes					
Laboratory, simulation Real-time analysis	Chemical processes	Biological processes	Mechanical Processes	Automation and use of ICT intervened programmes	
Production					
ERP, SAP, lean, green, zero waste technology	Footwear	Garments	Householdery	Auxiliaries	By-product
Marketing					
e-Commerce Supply chain management	International and domestic marketing	Product forecast	Global tanning and product sectors	Global trade in leather sector	GDP import and export details
Environmental Issues					
EMS, eAudit, ePortal	Suitable methods of disposing solid, liquid, gas wastages	Life cycle analysis, environmental management systems, environmental audit	Product recycle	Corporate social responsibility	Carbon footprint

7.1 Scope of KMS for Leather and Allied Processes and Industries

Central Leather Research Institute, the world's largest leather research institute (founded on 24 April 1948), has made an initiative with foresight to link technology system with both academy and industry. Today, CLRI is a central hub in Indian leather sector with direct roles in education, research, training, testing, designing, forecasting, planning, social empowerment, and leading in science and technology relating to leather. State-of-art facilities in CLRI support innovation in leather processing, creative designing of leather products, viz. leather garment, leather goods, footwear, and development of novel environmental technologies for leather sector. CLRI being one of the best educational institutes has got a huge collection of books on the leather and allied fields. The process of digital preservation enables the materials to be accessed through iPad, mobile, and kindle.

8 Building a Knowledge Management Portal

Knowledge management is a way to organize critical, reusable knowledge in a convenient, accessible way, available inside and outside the organization. It is also to pool in the individual creativity of all members of the community to generate a collective enterprise-wide innovation culture. The success of knowledge management system (KMS) fosters connecting right teams with right themes that is making availability of right information to right person at the right time conveniently. Enterprise resource planning (ERP) and KMS are complementary. KMS aims at improvements, innovations, and responsiveness, while ERP achieves transactional efficiency.

Advanced information retrieval technologies are becoming flexible with the "find more like this" features. Searchers across multiple sources, such as Internet, extranets, and a variety of file formats by a single application, are becoming common. Document summarizing and automatic communication to users are possible. All these make KMS efforts more powerful and rewarding. There are several IT tools available for information gathering and knowledge generation. Internet, intranet, extranet in an organization facilitate group working and collaboration. Artificial intelligence and neural networks help in building a strong knowledge-based system and find their application in imaging, pattern recognition, decision making, etc. e-mail, teleconference, videoconference, chat, etc., help in communication and collaboration systems. Content management software helps in creating multimedia content which can be deployed for e-content and e-Learning.

The twenty-first century demands better and greener technologies, so there is a need to know the issues and concerns regarding environmental pollution in an industry. Developing an ePortal on environmental issues will help an organization to increase its efficiency in natural resource management and diminish wastes and emissions. An e-Learning portal is a website that offers learners or organizations consolidated access to learning and training resources from multiple sources. The portal created will be according to the norms put forward by Sharable Content Object Reference Model (SCORM). SCORM defines a specific way of constructing learning management systems (LMS) so that they work well with other SCORM compliant systems, so it is important that the portal created meets the specifications put up by SCORM.

This proposed portal focuses on the various principles of environmental management, its goals and the need for sustainable development, the policies and legal aspects of environmental management. It also focuses on the impact that a

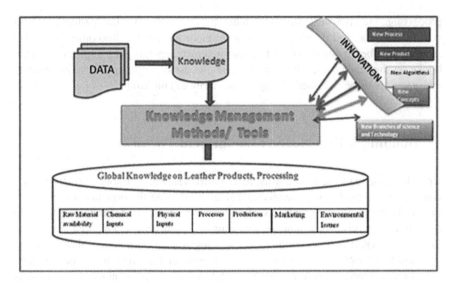

Fig. 13 Certain components of KM portal for leather and allied fields [11]

particular operation will have on the environment and the environmental clearance procedure in India. It deals with the carbon footprints and an attempt to calculate the carbon footprint of various processes in the leather industry mainly focusing on chrome tanning and vegetable tanning processes.

This portal once fully developed will give an insight into the environmental auditing focusing on the ISO 14001 requirements, the audit tools and technology, and the certification process. The various environmental management techniques like environmental monitoring, environmental modelling, sensitivity analysis, application of remote sensing and GIS in environmental management, and eco-mapping will be focused. Also, an attempt to find a relation between the gross domestic product and the carbon footprint of a particular industry and its products will be made. A safety manual on leather processing will serve as a knowledge resource for students and workers.

The green industry agenda covers the greening of industries, under which all industries continuously improve their resource productivity and environmental performance. Building an Enterprise Mission Management Portal—a knowledge park (Fig. 13) is essential for the success of the industry.

8.1 Tangible Outcomes

Figures 14, 15, and 16 illustrate the creation of Big Data, text, and image analytics search facility and accessing the data through iPad and mobile.

Fig. 14 A snapshot of display of eBooks on leather processing by Readium, Ipad [23]

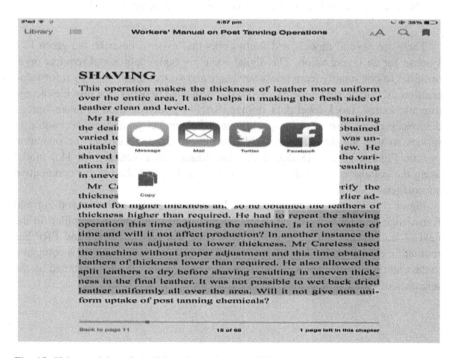

Fig. 15 Using social media collaborative endeavours [24]

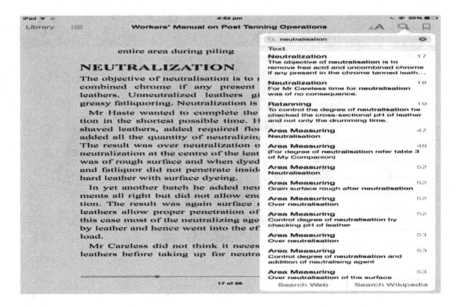

Fig. 16 Search facility in Big Data: data mining for the unstructured data [23]

There are several models and frameworks that exist to describe the green ICT systems for an organization. The digital economy both enables and requires organizations to continually learn new knowledge and systematically deploy it for value creation. A storage area network (SAN) is a dedicated network that provides access to consolidated, block-level data storage. SAN is primarily used to make storage devices, such as disc arrays, tape libraries, and optical jukeboxes, accessible to servers so that the devices appear like locally attached devices to the operating system. Figures 17 and 18 depict the visualization of e-Tannery for I4.0 model using the collaborative model for knowledge transfer and sharing for competitive advantages.

Figure 18 Portrays a framework for Innovation Centre with Networking and Knowledge transfer between academic institutes and industries a key pillar. In the field of leather processing, applying knowledge management study and ERP has resulted in cleaner technology (Fig. 19) and increase in profit margins as compared to the traditional leather processing. In-plant technologies [25] demonstrated by the scientists of CLRI to tanneries are listed below.

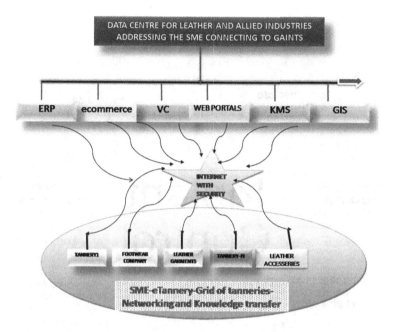

Fig. 17 Data centre for leather industry, industry 4.0 [11]

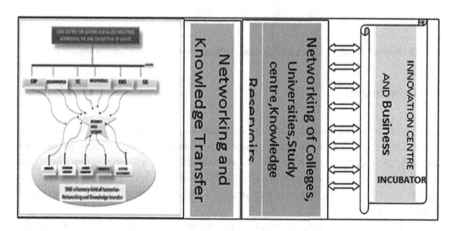

Fig. 18 A framework for innovation centre

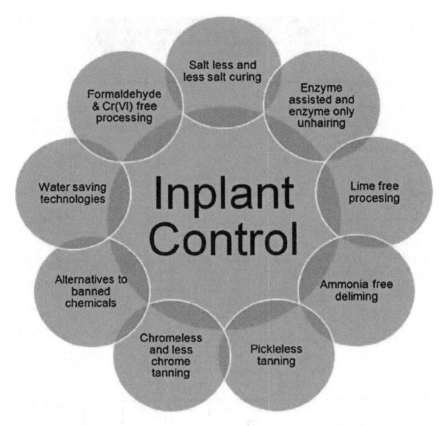

Fig. 19 In-plant innovative technologies demonstrated to tanneries. *Source* www.clri.org

9   Conclusion

Digitalization of an industry activity and knowledge pool that resides in an academic institute will drive quantum leaps in performance. The synergy between the industry and academia will accelerate globalization. It will strengthen the digital relationships with the empowered customers and also will focus on people and culture to drive transformation. Data analytics and digital trust are the foundation of industry 4.0.

In this chapter, a way to implement industry 4.0 standards for the leather industry using ICT tools and the existing knowledge base on cleaner and greener technology of leather processing enabling a digital eco-business models and decision support system has been discussed. The steps described in the flowchart 1 can be followed by other industries as well. This will increase manufacturing productivity, shift economics, foster industrial growth, and modify the profile of the workforce—ultimately changing the competitiveness of companies. In future, advanced implementation of industry 4.0 will become a qualifier to compete and it ensures digital ecosystem to be established throughout the value chain. A new business model can be generated using digitalization that will enable some new revenue opportunities. Synergy between industry and academia will result in innovation and prepares the launch pad for the country to gain global leadership.

Acknowledgements The author wishes to thank the Director, Central Leather Research Institute, Adyar, Chennai, for the motivation given to carry out this work. This work is an extension of the work presented at 10th Asia International Conference of Leather Science and Technology (AICLST) by the author titled "eGov_SD_eStem for Leather Industry" at Okayama, Japan, during Nov 2014. Author wishes to thank the Workshop Coordinator Dr. Usha Mujoo Munshi and Prof. Dolly Arora, Indian Institute of Public Administration and Department of Science and Technology, New Delhi, for providing a very detailed course work on the topic knowledge management and knowledge sharing in an workshop organized during Sep 6–12, 2015.

References

1. DIPP. (2012). Department of Industrial Policy & Promotion. Report of working group on leather and leather products: Twelfth five year plan period (2012–17). Retrieved from http://planningcommission.nic.in/aboutus/committee/wrkgrp12/wg_leath0203.pdf.
2. IICCI. (2008). The Indian leather industry, the Indo-Italian Chamber of Commerce & Industry. Short market overviews from https://smallb.sidbi.in/sites/default/files/knowledge_base/TheIndianLeatherIndustry.pdf.
3. DIPP. (2016). Department of Industrial Policy & Promotion. Leather sector achievement report 2016 from http://www.makeinindia.com/article/-/v/leather-sector-achievement-report.
4. DIPP. (2016). Department of Industrial Policy & Promotion. Leather Sector, Achievement reports, make in India, from http://www.makeinindia.com/documents/10281/114126/Leather+Sector+-+Achievement+Report.pdf.
5. Thyagarajan, G., Srinivasan, A. V., Amudeswari, A. (1994). Indian leather 2010, a technology industry and trade forecast, Directions for the future, (P:I-221). The Central Leather Research Institute, Adyar, Madras.
6. Design for the Environment Life-Cycle Assessments from https://www.epa.gov/saferchoice/design-environment-life-cycle-assessments.
7. Small and Medium-Sized enterprises in Economic Development: The UNIDO experience, Sarwar Hobohm, Industrial Development Officer, UNIDO, Vienna.
8. Responsible production a framework for chemical hazard management for small and medium sized enterprises incorporating best practice from safer production. APELL and Corporate Social Responsibility from http://www.unep.fr/scp/xsp/saferprod/pdf/Responsible_Production_Framework.pdf.
9. Global Industry 4.0 Survey. (2016). What we mean by Industry 4.0/survey key findings/blueprint for digital success from https://www.pwc.com/gx/en/industries/industries-4.0/landing-page/industry-4.0-building-your-digital-enterprise-april-2016.pdf.
10. Anantharaman, L. (2011). Study of new technologies of information and communication (TIC) in tanning industry. *Leather News India, 4*(4), (2013), XXXI IULTCS (International Union of Leather Technologists and Chemists Societies) CONGRESS, Valencia, Sep 2011.
11. Anantharaman, L. (2012). Knowledge management in an organization, industry-academy partnership in the context of leather and allied fields. In *IEEE Xplore Digital Library-2012 15th International Conference on Interactive Collaborative Learning* (ICL). https://doi.org/10.1109/ICL.2012.6402084.

12. "The enGauge 21st-Century Skills" is adapted from materials provided by NCREL, North Central Regional Education Laboratory from http://www.ncrel.org/engauge/skills/21skills. htm.
13. SCORM Overview from http://adlnet.gov/adl-research/scorm/.
14. Anantharaman, L. (2012) *Knowledge management and elearning. IEEE Xplore Digital Library-2012. In 15th International Conference on Interactive Collaborative Learning* (ICL). doi:https://doi.org/10.1109/ICL.2012.6402171.
15. Nixon, R. (2015). PHP Functions and Objects (4th ed.). In T O'Reilly (ed.), *Learning PHP MySQL & Javascript with Jquery, CSS and HTML5.* USA.
16. Gilmore, W. J. (2011). *Web development. In: A programmer's introduction to PHP 4.0.* US: Apress,.
17. The environmental impact of the animal product processing industries from http://www.fao. org/WAIRDOCS/LEAD/X6114E/x6114e05.htm#b4-3.2.1.%20Solid%20waste.
18. INCCA. (2007). Indian Network for Climate Change Assessment. Overview of GHG emissions from the energy sector. India: greenhouse gas emissions 2007, Ministry of Environment and Forests Government of India from http://www.moef.nic.in/downloads/ public-information/Report_INCCA.pdf.
19. Environmental technology from http://www.clri.org/TechnologyDetails.aspx.
20. Liu, H., Sun, J., Zhu, Z. (2008). Study on Process reengineering of Transportation Planning. In *International Conference on Intelligent Computation Technology and Automation* (ICICTA), P494–497. https://doi.org/10.1109/ICICTA.2008.230.
21. Ramasami, T., Krishnama Naidu, B., Chandramouli, et al. (2005). Report of all India survey on raw hides & skins. Chapter VI, P-190, Central Leather Research Institute.
22. R&D Activities. www.clri.org.
23. Anantharaman L (2015) Creation of knowledge portal using latest IT tools and mobile content delivery system: Work in progress, IEEE Xplore Digital Library. https://doi.org/10.1109/ TALE.2015.7386053.
24. Ramasami T., Workers manual on post tanning operation. Central Leather Research Institute, Adyar, Chennai.
25. The Leather Post, Vol. 1, 11th ed., April 2017. www.clri.org.

Applying Big Data Analytics in Governance to Achieve Sustainable Development Goals (SDGs) in India

Charru Malhotra, Rashmi Anand and Shauryavir Singh

1 Introduction

The proliferation, customization, and amalgamation of technology in the operationalization of the government's strategies, through the implementation of e-governance, have proven itself to successfully support governance tasks and handle the humungous data generated in the process thereof. As a means to this end, e-governance acts as a foundation for the government to promote good governance in three basic ways: by increasing transparency, accountability and maximizing the use of available information; by facilitating accurate decision making through effective public participation; and by ensuring the efficient delivery of public goods and services [1].

In the same vein, some of the apparent advantages of big data analytics, already being savored by corporate organizations, also hold true for the public sector agencies. These advantages are cost reduction, faster and better decision making, and insightful design of new products and services. For instance, Big Data techniques can facilitate storage of large amount of data and bring cost advantages to the public organizations. Hence, it is not surprising that big data analytics has now been acknowledged as one of the prime focus areas for governance to catalyze

Charru Malhotra, Anand R., & Singh. S. (2017). Applying Big Data Analytics in Governance to Achieve Sustainable Development Goals (SDGs) in India. In U. M Munshi., & N. Verma (Eds), *India In Data Science Landscape Towards Research Standards and Protocols (due for publication in 2017)*. Singapore, SG: Springer International Publishing AG.

C. Malhotra (✉) · R. Anand · S. Singh
Institute of Public Administration (IIPA), New Delhi, India
e-mail: charrumalhotra.iipa@gov.in

© Springer Nature Singapore Pte Ltd. 2018
U. M. Munshi and N. Verma (eds.), *Data Science Landscape*,
Studies in Big Data 38, https://doi.org/10.1007/978-981-10-7515-5_19

social, economic development of the societies. In a more advanced scenario, which too is not very distant, big data analytics coupled with emerging trends in artificial intelligence[1] have together been touted as the future of governance.

The United Nations has already recognized the potential of big data analytics to enhance evidence-based decision making at global, national, and local levels and to drive the implementation of Sustainable Development Goals (SDGs).[2] Even in the context of India, the journey had started back around the year 2010 onwards, Indian elections of the year 2014 being a case in point, where the political leaders had used to their advantage, big data analytics for citizens' sentiment analysis.[3] Now it needs to culminate at the point where the overall development agendas and better public service delivery experiences may be met successfully by the governance agencies by data mining complex, unstructured, imperfect data collected from heterogeneous sources including mobile and cloud that are being used by the citizens. Therefore, an effective analytical approach is the need of the hour to decipher the challenges of big data and analytics in supporting the governmental processes in the economy. The main aim of this paper is to illustrate the need, implementation strategy, and related issues of availing big data analytics in governance to achieve Sustainable Development Goals for India by the year 2030, keeping in sight its existing Digital India program.

1.1 Nature of the Study

Data is growing faster than ever before. With the increase in storage capabilities and methods of data collection, huge amounts of data have become easily available, which need to be discerned systematically by the public agencies to gauge actionable insights in governance processes to the advantage of the citizenry. This study first encapsulates the theoretical ruminations on the emerging topic of big data analytics with special reference to governance and subsequently maps governance targets established by SDGs with corresponding benefits offered by big data analytics in the process of their successful achievement. Then, a brief sketch of the seventeen Sustainable Development Goals (SDGs) mapped with related digital initiatives and existing mission mode projects (MMPs) of the Digital India program is provided. After illustrating this one-to-one correspondence of SDGs with a range of big data analytical techniques as well as with Digital India initiatives, the study concludes by summarizing the related challenges and concerns of adapting big data analytics in governance.

---

[1]Artificial intelligence is the simulation of human intelligence processes by machines, especially computer systems. These processes include learning, reasoning, and self-correction.

[2]http://sdg.iisd.org/news/un-highlights-role-of-big-data-in-achieving-big-impact-for-development/; Retrieved on July 18, 2017.

[3]http://dataconomy.com/2014/11/indian-government-using-big-data-to-revolutionise-democracy/; Retrieved on July 13, 2017.

1.2 Organization of the Paper

Section 1 of this paper presents a summarized view of literature discussions on big data and big data analytics with special reference to public service delivery and governance concerns and then describes the Sustainable Development Goals (SDGs) laid out by the United Nations. Specifically, the first subsection of this study (Sect. 2.1) presents the literature review related to big data and data analytics for governance, then explores the background and genesis of Sustainable Development Goals (Sect. 2.2), and attempts to map the application of big data analytics particularly in the context of Sustainable Development Goals (Sect. 2.3). Section 3 highlights research gaps and objectives for relating big data analytics to the public domain, especially in the Indian context. The next section (Sect. 4) summarizes some possibilities of applying techniques associated with big data analytics (such as sentiments analysis, predictive analysis, data fusion, and integration) to an assortment of sources of big data (such as mobile, sensors, and RFID devices) in the governance context. The same section also attempts to understand the prevalent ecosystem of Indian governance with reference to SDGs being mapped to initiatives under the Digital India program. In the last part of Sect. 4.1, the study presents a very crisp delineation of challenges associated with the application of big data analytics in governance. Section 5 summarizes the core discussions of the study as concluding remarks.

2 An Overview: Big Data Analytics, SDGs and Achieving 2030 Agenda Using Big Data Analytics

Big data analytics is extremely valuable to increase process productivity, opening up immense opportunities to make great progress in various aspects of governance. In particular, it supports the greater demand for transparency in governance and provides the evidence base for effective decision making. It, therefore, holds the power to contribute to the achievement and to monitor the progress of SDGs in various ways.

2.1 Literature Review—Big Data and Big Data Analytics in Governance

In the present times, data is being generated at a rapid pace, due to the increased use of technology such as the Internet, digital devices, mobile phones, RFID sensors, and social networking sites. Actually, every second, the world produces more data than stored 20 years ago from all different data sets. In the year 2015, digital data from heterogonous sources was around 4.4 ZB (1 zettabyte = 10^{21} bytes) and it is estimated that until the year 2020, this accumulated digital data would grow to around 44 ZB (44 trillion gigabytes). Facebook users send on average 31.25 million messages and view 2.77 million videos every minute. A massive growth in video and photo data could be seen, where every minute up to three hundred hours of video are uploaded to YouTube alone. Smartphones play a crucial role in the collection of big data, as they are packed with sensors capable of amassing all kinds of data and not only the data created by the users themselves. The growing rate of such multimedia content and dissemination applications contribute to the increase in the quantity of data. In the year 2016, globally, 6.1 billion smartphone users were present and as per the report by Wireless Smartphone Strategies (WSS) services, global smartphone users will increase by 58% from the year 2016 to the year 2022. India too is leapfrogging in proliferation of digital devices and hence creating enormous digital data (Fig. 1).

With a stupendous growth of all such technologies and related social media services, a large volume of data is produced from different sources; this large amount of data is termed as Big Data. Data generated could be in the form of digital pictures, videos, posts, blogs, call logs, intelligent sensors, GPS signals, etc. This huge data crosses the limit of gigabytes and is classified into the category of Big Data. The term Big Data is largely characterized by the mix of the 4 V's—volume, velocity, variety, and veracity, a concept developed by Gartner in 2012. This

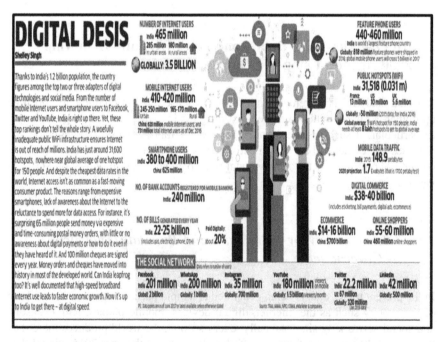

Fig. 1 Indian scenario of digital data. *Source* The Economics Times, Delhi, August 8, 2017, retrieved from http://epaperbeta.timesofindia.com/index.aspx?eid=31816&dt=20170808)

data-centric approach emphasizes not just the huge quantity of data and the increasing speed at which it is produced nowadays but also the range of the formats being processed [2]. The (US) National Institute of Science and Technology has defined big data as data, which 'exceed(s) the capacity or capability of current or conventional methods and systems.'

As is evident from the preceding section as well as from the context of the Indian illustration (Fig. 1), every second, more and more data is being created from heterogeneous sources and needs to be analyzed in an integrated manner in order to extract maximum value out of it, which is enabled by the science of big data analytics. Various authors have defined the term data analytics and made its interpretation in different notations for their studies. In a very basic form, big data analytics is a collection of different types of tools, including those based on predictive analytics, data mining, statistics, artificial intelligence, and natural language processing (Transforming Data with Intelligence—TDWI report, 2013). The main goal of big data analytics is to help organizations make better business decisions and forecast future trends.

With the development of e-government initiatives in India, the transactional data is growing exponentially. It has the ability to change the landscape of e-governance projects; like how data is generated, maintained, analyzed, and used for taking future decisions based on the results generated through analytics [3]. These aspects of big data can affect reliability of data, and sometimes handling and managing data become tough due to variability. Several authors (for instance, [4]) have specifically highlighted several big data challenges, such as difficulties in data capture, data storage, and data analysis and data visualization. A summary of select literature (Table 1) explored on the subject area presents different perspectives of using big data analytics in businesses and for government.

The review of literature (Table 1) clearly affirms that big data analytics is paving the way to transform how governments are using information technology to gain insight from their data repositories to make informed decisions. However, it needs to be also understood that such huge and heterogeneous data coming from several sources often arrives at high speed and is of such a diverse nature that for a particular problem domain, at a given point of time, it cannot be efficiently processed or utilized using current/existing/established/traditional technologies and techniques.

2.2 Sustainable Development Goals (SDGs)

Shared indicators and statistical frameworks always help entities to gauge how they are doing toward fulfillment of a particular objective, both internally as well as in comparison with others. On the same line of thought, it is being realized that for a government to plan and monitor the impact of its policies, it too must be able to benchmark data and measure year-on-year progress on certain development indicators. With this understanding, Sustainable Development Goals (SDGs), also

Table 1 Literature review on application of big data analytics

S. No.	Paper title	Scope of the research	Key issues dealt
1	Perspective on big data and big data analytics [5]	This paper defines concept of big data and stresses on the importance of big data analytics	Big data analytics software has been discussed in details with special reference to the challenges
2	Big data analytics: a literature review paper [6]	This paper aims to analyze some of the analytics methods and tools, as well as the opportunities, provided by the application of big data analytics in various decision domains	The main focus of this review paper has been to study the literature with prime focus on improving performance of supply chain with the help of big data analytics tool
3	Big data analysis and its need for effective e-governance [7]	This paper focuses on the impact of big data analysis for e-governance	More specifically, use of big data has been covered in healthcare industry especially for drug discovery
4	Big data analytics: challenges and applications for text, audio, video, and social media data [8]	In this paper, issues, challenges, and application of statistical data, text, audio, video, sensor, and biometric data are discussed	Social media data analytics and related issues and application areas have been put forth
5	Data-intensive applications, challenges, techniques, and technologies: a survey on Big data [4]	This paper is aimed to demonstrate a close-up view about big data, including big data applications, big data opportunities and challenges, as well as the state-of-the-art techniques and technologies we currently adopt to deal with the big data problems	– Big data is extremely valuable in increasing productivity of organizations – Big data brings many challenges, such as difficulties in data capture, data storage, and data analysis and data visualization
6	Role of big data analytics in analyzing e-governance projects [3]	This paper focuses on role of big data analytics in e-governance projects	Benefits of using big data analytics for e-governance projects have been elaborated
7	Big data, open government, and e-government: issues, policies, and recommendations [9]	This paper focuses on the potential and role of Big and Open Data in transforming e-government services. It further aims to decode how these technologies could be helpful in bringing	– Big data can help in identifying citizen requirements in real time. It has the capability to provide solutions to problems with efficiency and effectiveness

(continued)

Table 1 (continued)

S. No.	Paper title	Scope of the research	Key issues dealt
		openness and transparency in the government sector	– The paper presents Big and Open Data can foster collaboration, create real-time solutions to challenges in agriculture, health, transportation, retailing, tourism, etc.
8	Big data analytics and firm performance: Effects of dynamic capabilities [10]	The study examines the direct effects of big data analytics Capability (BDAC) on firm performance	While proposing a big data analytics capability using BDAC model, the study affirms that the application of big data analytics is very context-specific and shall vary from application to application
9	Big data analytics for supply chain management: A literature review and research agenda [11]	The main objective of this paper is to provide a literature review of big data analytics for supply chain management	– A review of articles within SCOPUS database related to the topics was done – The focus was on the use of data analytics for better decision making in supply chain management
10	An Anatomization of *Aadhaar* card data set—A big data challenge [12]	This paper provides an example for total number of *Aadhaar* cards approved by the state, rejected by the state, total number of Aadhaar card applicants by gender, etc.	Different challenges encountered in analyzing and extracting meaningful information from *Aadhaar* data in Indian context
11	Big data applications in real-time traffic operation and safety monitoring and improvement on urban expressways [13]	This paper focuses on the viability of a proactive real-time traffic monitoring strategy, that helps in evaluating operation and safety simultaneously	– The big data generated by the intelligent transportation system (ITS) is useful for proactive traffic management – Data mining and Bayesian inference techniques are useful in implementing real-time traffic crash prediction models
12	Data analytics for rural development [14]	This paper provides conceptual framework for the application of data analytics in enhancing	– The success of big data for rural development depends on the participation and support

(continued)

Table 1 (continued)

S. No.	Paper title	Scope of the research	Key issues dealt
		rural development by supporting different sectors such as agriculture, banking, governance, and health care	of both public and private bodies – The solutions that can provide real-time analysis of needs of the rural people should be developed
13	Big data proposes an innovative concept for contesting elections in Indian Subcontinent [15]	This paper focuses on the use of big data in the process of conducting elections that would help the political parties in canvassing and targeting voters. On the other hand, Big data technique would also help the electorates to appoint an efficient representative from their constituency	Big data is an emerging technology, so there is a need to develop expertise and for creating small-scale prototypes that can quickly be used for testing to demonstrate their correctness, matching with goals
14	Role of big data analytics in rural health care—a step towards Svasth Bharath [16]	This paper is aimed to present the reforms in the healthcare sector and presents a view on how government can harness innovations in the big data analytics to improve the rural healthcare system	– The findings of the paper identify that there is massive shortage of proper healthcare facilities in rural India – Critical computing and analytical ability of big data in processing huge volumes of transactional data in real-time situations can be used to turn the dream of Svasth Bharath (Healthy India) into reality
15	Urban planning and building smart cities based on the Internet of Things using big data analytics [17]	This paper proposes an IoT-based system consisting of various types of sensor deployment, viz. smart home sensors, vehicular networking, weather and water sensors, smart parking sensors, and surveillance objects for smart city development and urban planning using big data analytics	– The type of system proposed in this paper with full functionality does not currently exist. So, the results demonstrated by proposed system are more scalable and efficient than existing systems – Historical offline data is analyzed and used for urban planning and future development of cities

known as the global goals, have been built upon the Millennium Development Goals (MDGs) and been adopted on September 25, 2015.

One of the most significant dimensions of the SDGs has been the emphasis on using scientifically grounded indicators for each of the SDGs that would help to monitor progress, implement strategies, allocate resources, and increase the accountability of stakeholders involved. With this underlying motivation, SDGs encompass seventeen (17) goals covering one hundred and sixty-nine (169) targets that are to be achieved by the year 2030 for attaining sustainable development. These seventeen SDGs are titled—no poverty, zero hunger, good health, and well-being, quality education, gender equality, clean water and sanitation, affordable and clean energy, decent work and economic growth, industry, innovation and infrastructure, reduced inequalities, sustainable cities and communities, responsible consumption and production, climate action, life below water, life on land, and peace, justice and strong institutions. These goals seek to end all forms of poverty, fight inequalities, tackle climate change, and address a range of social needs like education, health, social protection, and job opportunities by the year 2030. The UNDP rightfully identifies SDGs as 'a universal call to action to end poverty, protect the planet, and ensure that all people enjoy peace and prosperity' and all the participating countries are committed to achieve the same.

2.3 Application of Big Data Analytics for Achieving SDGs

As already mentioned in the earlier section, United Nations is already recommending the need for partnerships to leverage the potential of the 'data ecosystem,' including big data applications across the sectors, in a holistic way. Successful achievement of SDGs by the year 2030 would require leveraging emerging technologies, including big data analytics, for supporting the related processes of governance, some of which are illustrated in the present study (Table 2), only as an example and not in any way as an exhaustive list.

This mapping between SDGs and big data analytics clearly highlights a need to demystify technological possibilities, potential avenues, policy opportunities, and related challenges that might emerge while engaging with techniques and technologies related to big data and big data analytics at the local level. To identify these opportunities and challenges, several innovative models, frameworks, and virgin possibilities need to be understood to resolve systemic governance issues, to achieve sustainable development in the country.

Table 2 Application of big data analytics for achieving SDGs

S. No.	SDG goals	Some suggested application of big data analytics
1	No poverty	Due to the telecom revolution in India and high penetration of mobile phone services, even in majority of the rural areas, it is much easier to generate poverty maps through tele-density records giving an insight into the communication and mobility patterns of the affected population. With the construction of such maps, diagnosis of the levels of poverty in various areas and taking steps to counter it can happen at a decentralized local level. Through the high resolution of geographical insights provided by these maps in the living standards, employment status, access to healthcare, education, etc., it is much easier for governance agencies to take effective measures to eradicate the various facets of poverty
2	Zero hunger	Through precision agriculture,[a] governance agencies can use big data analytics on the data collected from various high-tech sources such as aerial images, embedded local sensors, and weather departments to develop solutions at a much more finely geographical resolution in farmlands rather than providing a common solution that works well for most but not best for any
3	Good health and well-being	Using data collected from mobile phone mobility patterns can predict and then prevent the spread of infectious diseases
4	Quality education	The analysis of Big Data collected from the education sector can reveal various flaws in the education system of the country, such as decoding the student dropout rates specific to particular geographies
5	Gender equality	Analysis of financial transactions can reveal the spending patterns of men and women; further these patterns can reveal the differences in financial and economic well-being, independence, growth, etc. between men and women and thereby help to plug-in the gaps to ensure a balanced growth of the economy
6	Clean water and sanitation	Sensors connected to water pumps can track access to clean water
7	Affordable and clean energy	Smart metering allows utility companies to increase or restrict the flow of electricity, gas or water to reduce waste, and ensure adequate supply at peak periods
8	Decent work and economic growth	Patterns generated through analysis of—connected devices, sensors can provide indicators such as remittances and GDP

(continued)

Table 2 (continued)

S. No.	SDG goals	Some suggested application of big data analytics
9	Industry, innovation, and infrastructure	Data from GPS devices may be used for traffic control and to improve public transport
10	Reduce inequality	Speech-to-text analytics on local radio content can reveal discrimination concerns and support policy response
11	Sustainable cities and communities	Satellite remote sensing can track encroachment on public land or spaces such as parks and forests
12	Responsible consumption and production	Online search patterns or e-commerce transactions can reveal the pace of transition to energy efficient products
13	Climate action	Combining satellite imagery, crowd-sourced witness accounts and open data can help track deforestation
14	Life below water	Maritime vessel tracking data can reveal illegal, unregulated, and unreported fishing activities
15	Life on land	Social media monitoring can support disaster management with real time information on victim's location
16	Peace, justice, and strong institutions	Sentiment analysis of social media can reveal public opinion on effective governance, public service delivery, or human rights
17	Partnership for the goals	Partnerships between industry, academia, and civil society to enable the combining of statistics, mobile and internet data for providing a better and real-time understanding of today's hyper-connected world

Big data applications w.r.t to each SDG goal have been culled out and analyzed from UN Global Pulse report, 2016; this is just an illustrative list and not at all an exhaustive one
[a]http://cio.economictimes.indiatimes.com/dobig/news/detail/1623; Retrieved on July 26, 2017

3 Research Gaps and Objectives

Based on the literature review on various aspects of big data, analytics and their specific application in the public sector, it has been understood that several governance process improvement opportunities arise in sectors such as transport, health (Sect. 2.1) and several more remain untapped when viewed from the perspective of SDGs (Sect. 2.3). The significance of achieving these SDGs as mandatory targets is very high for a democratic country like India that aspires to achieve rapid development in economic, social, and environmental sectors.

Based on this theoretical premise, a cursory mapping is undertaken between the ambitious Digital India program and the seventeen SDGs (Sect. 4.1). This relationship underscores the fact that Indian governance scenario is still 'missing the forest for the trees.' Several quiescent SDG targets beseech holistic digital systems to address the governance challenges of the country. Thus, the next section of the study (Sect. 3) attempts to understand the application of big data analytics in the governance realm, with special reference to India and its Digital India endeavor.

4 Employing Big Data Analytics in Governance: Addressing the Concerns in India

Good quality of life and an equitable happiness index for the state is what each governance delivery agency aims for. To do so, real-time data is imperative. Data analytics could be used for taking smart decisions in the ever-changing and dynamic environment. There is a lot of data being generated by different sources, viz. social media, sensors, RFID devices, government departments, and so on. If the data generated is properly analyzed through various techniques of big data analytics, then it could provide great support to effective decision making in governance (Fig. 2).

The citizens of the present era are more demanding and more connected than ever before due to the proliferation of multiple interfaces to which he/she is exposed through emerging technologies. In addition, at each touch point of these interfaces, the citizen leaves behind a 'digital footprint on the sands of devices.' It is for the delivery agents/agencies to gather these digital footprints and build up a holistic integrated image of their needs and aspirations. Big Data helps to 'collect' this data, and 'Data analytics' helps to 'mine' this data.

Keeping in view that the modern citizens are much smarter than before, the governance agencies must provide them exactly what they need and aspire for rather than force-fitting a 'one size fits all' solution. The application of big data

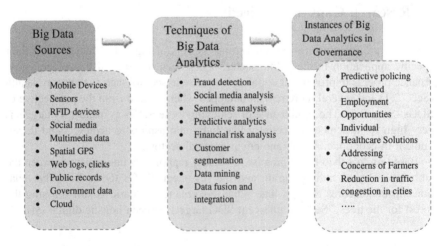

Fig. 2 Big data analytics and governance

The quiet digital revolution in Chandrapur

Field workers being trained on the Collect app, which was used to digitally collect data from every household.

Fig. 3 Digital revolution in Chandrapur. *Source* Livemint, August 26, 2017, retrieved from http://www.livemint.com/Sundayapp/QMbWi2VuPKfInCaNZnJbyL/The-quiet-digital-revolution-in-Chandrapur.html

analytics could be several—for instance, the steady stream of the collated baseline data from the grassroots can be helpful in designing customized public services and also for strategizing future grassroots innovations in the economies.

For example, a pioneering initiative of the Maharashtra government has data-mapped three blocks of the district at an unprecedented level. The final data from the baseline survey was then structured and analyzed to create village development plans, listing everything that needs to happen to transform each village into a model village (Fig. 3)

The baseline data hence collected from the governance tasks undertaken at the grassroots when combined with unstructured data created by user while using Internet, Social Media, mobile, cloud etc. can later help in not just continuous impact assessment of these tasks but can also help to uncover hidden patterns, unknown correlations, citizens' preferences, and other action able insights that can help public organizations make more-informed evidence-based governance decisions. Therefore, it is imperative that the contextualized aspirations of each of the citizens in a community of a constituency (city/village, municipality/urban local bodies-ULB/*Panchayat*) for all the states of the country may be mapped successfully by the delivery/governance agencies, by judicious application of big data analytics. Thus, keeping in sight the SDG targets, the prudent application of big data analytics can serve as the foundation of sustainable future efforts to create need-based public services and customized governance models, and further drive indigenous innovations to satiate the specific localized needs.

4.1 Indian Scenario: Digital India's Role in Achieving SDGs

With reference to the previous section providing a glimpse of applications of big data to address the challenges of SDGs, India too needs to assess its strategic opportunities for using big data to attain the SDG agenda by the year 2030. In the recent years, the Government of India (GoI) launched its umbrella program called Digital India that insists to shift from e-government to e-governance in a bid to 'transform India into a digitally empowered society and knowledge economy' (digitalindia.gov.in) to achieve economic, social, and environmental benefits. In the same spirit as SDGs, which call for building peaceful, inclusive, and well-governed societies, even Digital India framework resonates with the proposed SDG spirit by striving to work on three vision areas—Digital Infrastructure as a Utility to Every Citizen, Governance and Services on Demand, and Digital Empowerment of Citizens. To achieve these three vision areas, Digital India weaves together a large number of technologies with governance processes and services around nine pillars, which too resonate very well with seventeen SDG goals in one way or another. These nine pillars of Digital India include—broadband highways; universal access to mobile connectivity; public internet access program, e-governance: reforming government through technology; e-Kranti—electronic delivery of services, information for all, electronics manufacturing, information technology for jobs, and early harvest programs. In general, out of these nine pillars, four are about provision of Internet and access, which underpins a new prerogative for Indian citizens —'universal web access must be a human right,' and the others are more e-government/e-governance in nature, i.e., focus on providing governance and services on demand. Apart from some of these e-government/e-governance initiatives listed below (Table 3), Digital India also aims to spread digitally literacy among its marginalized population, especially the rural, the elderly, and women through several of its initiatives including National Digital Literacy Mission (NDLM). From the perspective of SDG attainment, this all-pervasive digital literacy in the country would ensure that majority of the citizens have the ability to create, locate, organize, understand, evaluate, and analyze 'data-islands' which would be co-created at the local level. In brief, it is possible to draw parallels of SDGs with various schemes/initiatives of the Government of India that, either directly or indirectly, aim to achieve the SDG targets (Table 3).

At present, there are several mission mode projects and related Digital India initiatives that attempt to achieve SDGs target wise (Table 3) for instance digital literacy mission would permit application of technology techniques like crowd sourcing to tangibly help government agencies in co-creating digital governance spaces. It is also understood that the contribution of Digital India in achieving the sustained development goals would be dependent on effective implementation of

Table 3 Mapping SDGs with some of the mission mode projects (MMPs) of Digital India initiatives and related e-initiatives in India

S. No.	Mapping SDGs with MMPs and Digital India initiatives
1	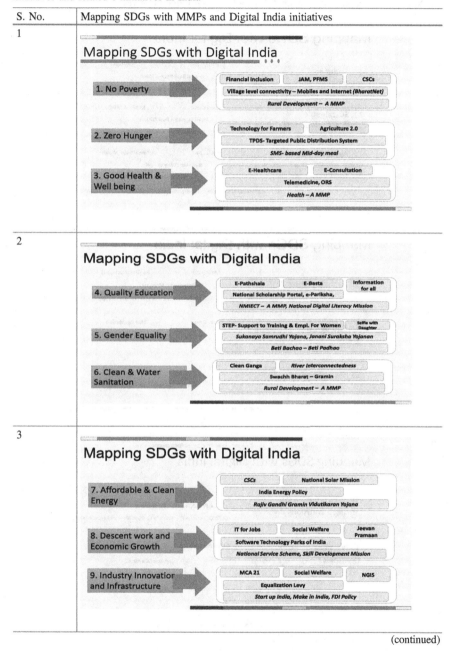
2	
3	

(continued)

Table 3 (continued)

S. No.	Mapping SDGs with MMPs and Digital India initiatives
4	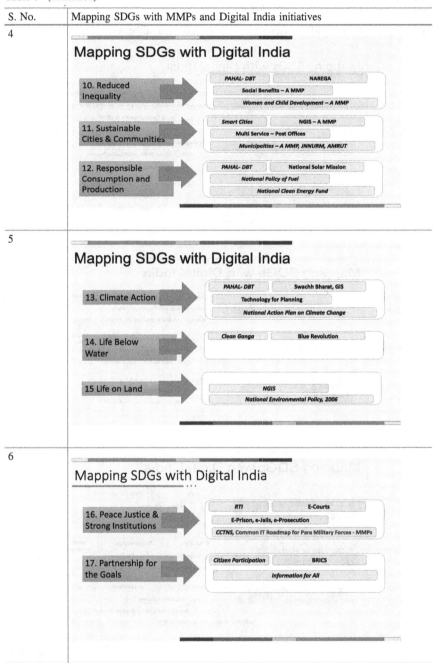
5	
6	

Note These are just instances of mapping of only some of the Digital India initiatives/MMPs and not an exhaustive delineation initiatives such as improved broadband penetration, creation of knowledge societies, local language applications and focus on e-health, e-education delivery model. Further, on deeper inspection of these Digital India initiatives, it is understood that majority of these initiatives have been designed as 'stand-alone' initiatives and not as one cohesive digital solution to address concerns of a particular governance sector. Multiple e-government/e-governance initiatives are operating in 'silos' and fail to holistically address issues and challenges of a particular sector in a cohesive manner. For example in education sector, (Table 3, S. No. 2) several initiatives such as e-*Basta* and e-*Pathshala* are sighted with information and functionality overlaps.

5 Concluding Remarks

Big data is not neutral, there are various challenges and risks involved in employing big data techniques/tools in governance, and these challenges need to be acknowledged and tackled through big data strategies. The most common types of challenges are infrastructure concerns, data deficit, data governance, reliability, quality, privacy, data anonymity, rightful ownership of data, data-price/value/cost for the citizens as well as for the user-agency and so on. Data privacy emerges as one of the key considerations, with various stakeholders calling for a unified legal and ethical approach. At present, more specifically, the capacities of government officials and citizens to analyze big data are also limited and must be strengthened in order to unleash the potential of big data in the policy cycle. Notwithstanding all such concerns, this study attempts to present various developments related to big data, data analytics, and their role in governmental processes toward achieving the Sustainable Development Goals, particularly in India. The present study also tie up with the approach of Indian governance toward achieving the sustainable goals through successful implementation of its Digital India program. After studying the subject area and issues highlighted in the paper, it is found that if big data is properly analyzed with good techniques and technologies, it can give rise to numerous ideas and will definitely help in decision making. It can be concluded that if the governance agencies use big data analytics for providing services to the citizens, then certainly it will not only give rise to more effective services but also help to incubate more responsive local governance systems. In the same breath, it is pertinent to point out that there would be various risks and data security issues with the big data technology-based applications. Therefore, toward its end, the study suggests that the government should focus on protecting the privacy rights of its citizens, *i.e.,* it is pertinent to protect the privacy of the citizen who yields the data or owns the data or is associated with the data. The upcoming decision of the apex court of India will be one of the most important legal decisions in the world this year. In this context, the constitutional controls on the corporates as well as the state have to be appropriately enforced. In a developing country like India, with the given lack of comprehensive and structured MIS data, their protection regulation and privacy law are the gray areas to improve upon, such initiatives could lead to issues such as ambiguity over who has data ownership, indiscriminate sharing of data across databases and data collected for one purpose could be used for other purposes.

Acknowledgements The authors jointly convey their gratitude to Mr. Trilok Chandra, ITS (Sr. Director, e-governance, Ministry of Electronics and Information Technology, Government of India, New Delhi) for providing us with the latest data on not just Digital India initiatives but also about the prevailing digital trends in the country. We are also thankful to Mr. Pushkaraj Gavali (Research Officer, e-governance, and Digital India, Indian Institute of Public Administration, New Delhi) and Dr. Vivek Soni (Research Associate, School of International Studies, Jawaharlal Nehru University, New Delhi) for their valuable inputs to stitch the whole paper as a united mass.

References

1. Magno, F. A., & Serafica, R. B. (2001). *Information technology for good governance*. Manila: Yuchengco Center for East Asia, De La Salle University.
2. Goes, P. B. (2014). Big Data and IS research. *MIS Quarterly, 38*(3).
3. Navdeep, P., Arora, M., & Sharma, N. (2016). Role of big data analytics in analyzing e-Governance projects. New trends in business and management: An international perspective.
4. Chen, C. L. P. & Zhang C. Y. (2014). Data-intensive applications, challenges, techniques and technologies: A survey on Big Data. *Information Sciences, 275*, 314–320.
5. Ularu, E. G., Puican, F. C., Apostu, A., & Velicanu, M., (2012). Perspective on big data and big data analytics.
6. Elgendy, N., & Elragal, A. (2014). Big data analytics: A literature review paper.
7. Agnihotri, N., & Sharma, A. K. (2015). Big data analysis and its need for effective e-governance. *International Journal of Innovations & Advancement in Computer Science, 4*.
8. Varma, J. P., Agrawal, S., Patel, B., & Patel, A. (2016). Big data analytics: Challenges and applications for text, audio, video, and social media data.
9. Bertot, J. C., Gorham, U., Jaeger, P. T. & Choi, H. (2014). Big Data, open government and e-government: issues, policies and recommendations. *ResearchGate*.
10. Wamba, S. F., Gunasekaran, A., Akter, S., Ren, S., Dubey, R. & Childe, S. J. (2016). Big data analytics and firm performance: Effects of dynamic capabilities. *Journal of Business Research*.
11. Wamba, S. F. & Akter, S. (2012). Big data analytics for supply chain management: A literature review and research agenda. *ResearchGate*.
12. Dayal, M., & Singh N. (2016). An anatomization of Aadhaar card data set—A big data challenge. In *International Conference on Computational Modeling and Security* (CMS 2016).
13. Shi, Q., & Abdel, M. (2015). Big Data applications in real-time traffic operation and safety monitoring and improvement on urban expressways. *ScienceDirect, Transportation Research Part C: Emerging Technologies, 58*, 380–394.
14. Peisker, A. & Dalai, S. (2015). Data analytics for rural development. *Indian Journal of Science and Technology, 8*(S4).
15. Jagdev, G., Singh, B. & Mann, M. (2015). Big Data proposes an innovative concept for contesting elections in Indian Subcontinent. *International Journal of Scientific and Technical Advancements*.

16. Kumar, M. N. & Manjula, R. (2014). Role of big data analytics in rural health care—A step towards Svasth Bharath. *International Journal of Computer Science and Information Technologies, 5*(6).

17. Rathore, M. M., Ahmad, A., Paul, A., & Rho, S. (2016). Urban planning and building smart cities based on the Internet of Things using big data analytics. *Science Direct-Computer Networks, 101,* 63–80.

Open Data: India's Initiative for Researchers, Research, and Innovation

Neeta Verma and Alka Mishra

1 Introduction: Exploring Potential of Open Data for Research

'Open data' describes the concept that information and data should be made available for everyone to access, reuse, and redistribute without any restrictions. Government organizations are repository of information in form of documents and data, and if these are made available following the principles for open data it can be referred to as 'Open Government Data'.[1]

For a country as diverse as India, the process of governance involves a lot of complexities. Deliberations and consultations with the various stakeholders within and outside the government are a key aspect of policy formulation in India. Given the important social and economic issues which the government has to deal with in India, it is more than imperative that government's decision while making process is an informed one, so that the success of its schemes and initiatives is ensured with a little margin for error. These schemes and initiatives which are targeted toward the citizens also require an inbuilt process to uphold accountability and transparency to ensuring that these schemes are actually benefitting the target users. Therefore, for an informed decision-making and transparent review of the government policies the essential ingredient is the availability of data.

---

[1]http://opendatahandbook.org/.

N. Verma (✉)
National Informatics Centre, New Delhi, India
e-mail: neeta@nic.in; neeta@gov.in

A. Mishra
MyGov & Open Government Data Division, National Informatics Centre,
New Delhi, India
e-mail: amishra@nic.in

© Springer Nature Singapore Pte Ltd. 2018
U. M. Munshi and N. Verma (eds.), *Data Science Landscape*,
Studies in Big Data 38, https://doi.org/10.1007/978-981-10-7515-5_20

2 Open Data: India's Initiatives

Government particularly in India collect process and generate a large amount of data in its day-to-day functioning, which are lying in silos and in files which are difficult to put to effective use [1]. Evidence-based planning is essential for socioeconomic development, and all this depends on availability of quality data. Asset and value potentials of data are widely recognized at all levels. Data collected or generated through public investments, when made publicly available and maintained over time, their potential value could be more fully realized. However, most of such data, which are non-sensitive in nature, remains inaccessible to citizens buried deep down in the government records and files. With open data, comprehensive statistical picture can be built and it also allows maximum use and reuse of data, which enhances the decision-making capability of the government and communities.

It is moved toward participatory governance[2] where:

- Government shares public datasets from its vast repository of data.
- Researchers and planners use it for analysis and planning.
- Developers develop new applications to bring data into context.
- Government and citizens collaborate.

3 Benefits of Open Government Data

The government by virtue of its size is the largest producer and consumer of the data. The government data that is placed in the public domain in a trusted environment stands benefitting many stakeholders.

Following are the benefits of Open Government Data:

- To support decision makers in government and public administration to make better and fact-based policy decisions and thus to increase government efficiency and effectiveness.
- To help citizens to better understand why and how decisions are made, which can help restore trust and can lead to better acceptance of policy decisions once enacted.
- To help citizens to hold their government and administration accountable, this can reduce corruption and mismanagement.
- To support and empower citizens to make informed decisions and engage with the government, thus enabling citizens to have a more active voice in society.

---

[2]Ten Principles for Opening up Government Information—https://sunlightfoundation.com/policy/documents/ten-open-data-principles/.

- To support governments, citizens, academia, and the private sector to work together and collaboratively find new answers to solve societal problems.
- To enhance public service delivery mechanism of the government through value addition to data. This will also enable the government to partner citizens in their daily work routine and engagement with government.
- To reduce the burden of locating information both for internal use as well as for responding to RTI applications and streamlining its own information gathering and processing procedures.
- To reduce the dependency on RTI applications for gathering information and details from government institutions, thus enabling effective use of government resources.
- To expose incorrect and outdated data, which the government itself is often not in an easy position to detect.
- To help citizens and the variety of civil society organizations in India currently working, despite difficulties, with government data.
- To help in innovation and economic growth.

4 Open Data Ecosystem: A Case for India

Recognizing the importance making open data available to its citizens for increased levels of proactive transparency and accountability and to promote higher level of public participation, the Government of India under the aegis of National Data Sharing and Accessibility Policy (NDSAP) [2] initiated Open Government Data (OGD) Platform India (https://data.gov.in) [3] to share government data with its citizens. This has built a foundation to create an open data ecosystem in the country. The platform has been set up and managed by the National Informatics Centre (NIC) under the Ministry of Electronics and Information Technology (MeitY), Government of India. Launched in October 2012, now has been loaded with more features and functionalities in its current version (Fig. 1).

Open Data Initiative of India paves the way for more efficient governance by facilitating easy access to plethora of government-owned datasets to public. Implementation of NDSAP [4] through OGD Platform has been increasing the usability and relevance of Open Government Data and sustaining the ecosystem around it. The OGD Platform has been developed and managed completely using open source stack. The platform has a rich mechanism for citizen engagement, which could help ministries/departments/organizations prioritize the release of government datasets. Besides, enabling citizens to express their need for specific datasets or apps, it also allows them seek clarification or information from nodal officers of participating government entities. It also acts as a knowledge-sharing platform and community participation through visualizations, APIs, APPs. Citizens are encouraged to contribute blogs, visualizations, info-graphics, and apps on various datasets through a separate community portal (https://community.data.gov.in) [5] (Fig. 2).

Fig. 1 A snapshot of Open Government Data (OGD) Portal of India

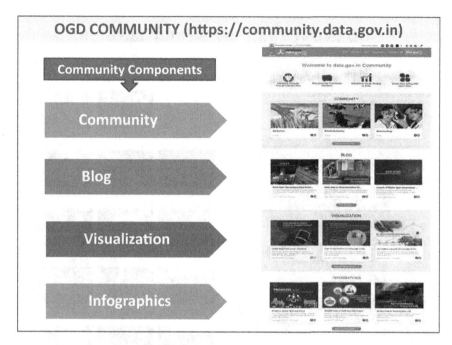

Fig. 2 A snapshot of OGD community

OGD demonstrated its potential to the App Developers' Community through various contests such as '12th Plan Hackathon,' 'In Pursuit of an Idea,' 'CMA Hackathon'. Out of which '#OpenDataApps Challenge' was launched in association with NASSCOM and 'Code for Honor 2014' was launched with Microsoft. A dedicated event portal (https://event.data.gov.in) has been created for management of workshops, hackathons, challenges, etc. OGD has been organizing events/apps, challenges/hackathons, and workshops/training/awareness sessions including state-level workshops [6] (Fig. 3).

Currently, OGD India has 100,000+ open data resources,[3] 4,130+ catalogs, 111 chief data officers. These dataset resources are 11.07 million times viewed and have been downloaded 4.55 million times.[4] More than 1094 visualizations have been prepared on these datasets. The platform provides dataset API's for the use of developers. Data Analytics is also facilitated (Fig. 4).

---

[3]https://data.gov.in/catalogs.

[4]https://data.gov.in/analytics.

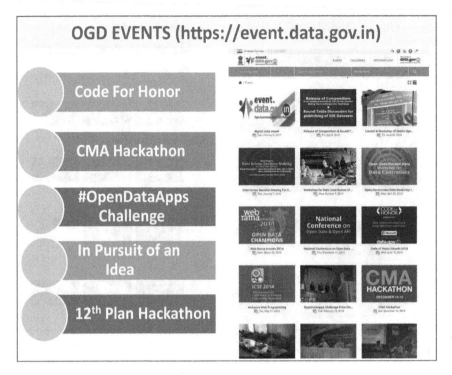

Fig. 3 OGD events

5 Government Open Data Use License

India has created and gazette notified to give a legal framework to the data users
wishing to use and build on top of public data. License also gives assurance of what
they legally can and can not do with the data both commercially and noncom-
mercially. The Journey of the Open Government Data Platform is given in (Fig. 5).

6 OGD for Research and Innovation

Apart from lots of benefits of open data to various stakeholders, public research and
private innovation opportunities expand with a policy of openness for upstream data
resources. Such data can substantially reduce unproductive barriers to interdisciplinary, interinstitutional, and international research. They enable data mining for
automated knowledge discovery in a growing sea of big data. Open data are
essential for the verification of research results and in generating broad trust in
them. They avoid many inefficiencies, such as the unnecessary duplication of
research and the identification of erroneous results. They promote more research

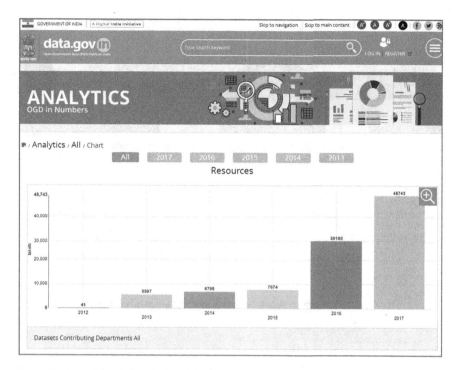

Fig. 4 A view of the data analytics of OGD

and new types of research. They permit the legal interoperability of data when multiple sources of data are combined for new knowledge.

Open data provides following benefits to researchers [7]:

- It allows **greater access to data** and open up possibilities to build upon and creates new research from publicly accessible data.
- Accelerates the pace of discovery and can be easily accessed and used to create a fuller picture.
- Helps ensure that breakthrough is not missed—someone else with a different perspective or analytical technique.
- Improves the integrity of the scientific and scholarly record.
- **Visibility of one's research** is enhanced.
- Working smarter and faster by **sharing and analyzing data** in real time.
- Allowing experiments to be **reproduced and verified** by anyone.
- Increasing researcher transparency and **reducing academic fraud**.
- Facilitating research across disciplines and fostering collaborations.
- Maximizing data reuse by **selecting formats** for efficient data extraction.
- Ensuring compliance with **funding agency mandates** and **journal publishing policies**.

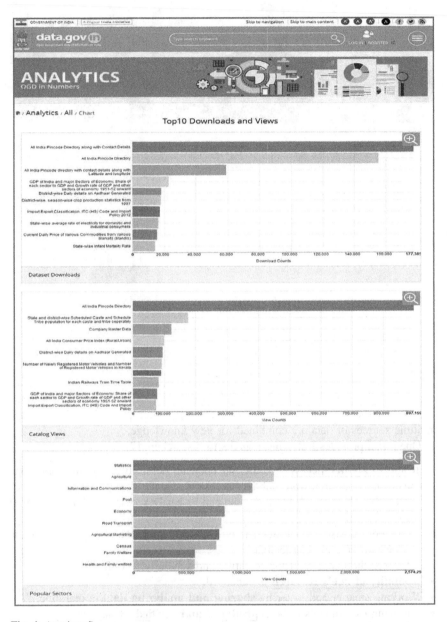

Fig. 4 (continued)

In most of the case, the data generated during the research becomes inaccessible or datasets are not properly archived in a manner that someone else can actually use it. Research budgets are shrinking, and number of researchers competing with each other is increasing. Moreover good datasets are needed to validate new techniques,

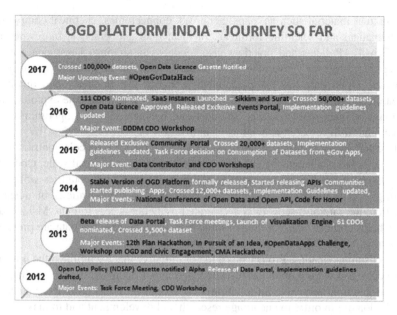

Fig. 5 OGD Platform India—milestone 2012–2017

calibrate computer models, and show comparative progress. So with shrinking budget and increasing research cost, availability of open datasets becomes critical and need of the hour which paves way for the timely and efficient use of data. Greater need has been felt to permanently archive open research data.

Broader access to scientific publications and data therefore helps to:

- Improve quality of results by building on previous research results.
- Increase efficiency by encouraging collaboration and avoiding duplication of effort.
- Fasten growth by speeding up innovation.
- Improved transparency of the scientific process by involving citizens and society.
- Improves visibility of research by reaching more people.
- Greater impact for science and society.

7 Open Data and Scientific Institutions

If we take the case of scientific institutions, openness is critical to the modern scientific enterprise. Science is increasingly data-driven and expensive, but access to scientific data is often subject to administrative, legal, and privacy regulations. It also requires adequate ICT infrastructure. Science open data may include non-textual material such as maps, genomes, connectomes, chemical compounds, mathematical and scientific formulae, medical data and practice, bioscience and biodiversity. Policies and practices also hinder data sharing. These concerns are being addressed by government to preserve and promote more openness in research. Open data has the potential to enhance the efficiency and quality of research by reducing the costs of data collection, by facilitating the exploitation of dormant or inaccessible data at low cost and by increasing the opportunities for collaboration in research as well as in innovation.

Greater access to research data can also help advance science's contribution to solving global challenges by enhancing access to data on a global scale. Open data can also be used to promote capacity building in developing countries while generating opportunities for scientific collaboration and innovation. In scientific research, the rate of discovery is accelerated by better access to data. Making data open helps combat 'data rot' and ensures that scientific research data are preserved over time.

The Human Genome Project was a major initiative that exemplified the power of open data. It was built upon the so-called Bermuda Principles, stipulating that: "All human genomic sequence information (…) should be freely available and in the public domain in order to encourage research and development and to maximise its benefit to society" [8].

8 Conclusion

Government being major player in developing governing policies for data access and sharing to foster greater access and use of scientific research can facilitate the sharing of data resulting from publicly funded research.

With the collaboration of government, researchers, and scientific institutions, proper steps can be taken to open up more scientific data for the use and benefit of society.

References

1. Compendium on Data Driven Decision Making—An Outcome of OGD workshop https://data.gov.in/sites/default/files/Compendium_Data_Driven_Decision_Making_NIC.pdf.
2. National Data Sharing and Accessibility Policy (NDSAP) of India—https://data.gov.in/sites/default/files/NDSAP.pdf.
3. Open Government Data (OGD) Portal India—https://data.gov.in.
4. Implementation Guidelines for National Data Sharing and Accessibility Policy (NDSAP) of India—https://data.gov.in/sites/default/files/NDSAP_Implementation_Guidelines_2.2.pdf.
5. Community Participation in OGD—https://community.data.gov.in.

6. Citizen Engagement initiatives and events of Open Government Data (OGD) Portal India—https://data.gov.in/events.
7. Benefits of Open Data to Researchers http://sco.library.emory.edu/research-data-management/open-data/benefits-research.html.
8. Open Data Meaning, Scope and Organizations—https://en.wikipedia.org/wiki/Open_data.

Data Management, Sharing and Services: Issues of Attitude Towards Data Citation and Role of Data Stakeholders

O. P. Wali

1 Introduction

With the advances in information technology, the incidences of allowing other researchers reuse or refer earlier data used by researchers, who use that data first time and publish, have grown overtime. There are opportunities as well as challenges to contribute as data providers which have affected the pace of growth in data sharing in public domain.

2 Issues of Attitude Towards Data Citation

There has been increase in interest to study various aspects of data sharing. One of the known and visible benefits for the contributors has been growth in citation. There are empirical studies, like in life sciences, to prove that paper citation correlates positives with availability of sample data for other researchers. It is also possible that other factors could drive citation score like reputation of journal, importance of the topic investigated, reputation of the investigator. But studies have proved the effect of data sharing as a significant variable to drive citation score which could be moderated by time lag. There is also evidence to show the growth of influence in the recent years which is indicative of more realisation among researchers to leverage this dimension for higher citation. This could in a way lead to a chain effect which means more and more data could be shared and combined to produce and add knowledge at a lesser cost.

O. P. Wali (✉)
Centre for International Trade in Technology, Indian Institute of Foreign Trade, New Delhi, India
e-mail: opwali@gmail.com

© Springer Nature Singapore Pte Ltd. 2018
U. M. Munshi and N. Verma (eds.), *Data Science Landscape*,
Studies in Big Data 38, https://doi.org/10.1007/978-981-10-7515-5_21

3 Data Reuse for Repurposing

The researchers can reuse earlier data or verify earlier results or improvise earlier findings, etc. However, data sharing behaviour is an aspect of human behaviour and behaviour could be function of individual traits, group's effects or belief system driven by the broader culture of place or geography. There is lack of desirable amount and width of rigorous research to investigate deeper into attitudinal issues affecting data sharing behaviour. The remedies to address behavioural tendencies are going to help speed up the growth of complete and authentic data availability vis-à-vis any mandatory standards. This in turn could lead to reduce cost and enhance reliability of findings about things in scientific research.

4 Do Share, Do Cite

When we use some shared data in our research work, it is our responsibility to cite the data source details and same goes for others who may use our shared data. Therefore, argument becomes simple. What we would like others to follow applies to us and also to whole scientific community. Why many of us still do not follow this could be because of various reasons, but the prime reason could be the attitude which needs to be addressed at much earlier stages of individual growth and development. Most of the other reasons can be addressed through sustained collaborative efforts across the global communities which include initiatives like this workshop and bring in desired normative discipline over a period of time.

5 Share and Collaborate

Sharing in scientific world and skills to manage collaboration can lead to a successful R&D endeavour at the state level. Policies and diplomatic efforts to foster collaboration for technology developments built on exchanges or collaborations for basic and applied R&D efforts succeed only when scientists intend honest collaboration and sharing. Lot of efforts have been made by countries like China and Korea in recent past to enhance research collaboration, research visibility and diversity of research to spur new knowledge and recognition of local research. Sharing attitude is at the heart of this deliberate effort to collaborate.

6 Data Citation (DC) Space

With so much of growth and progress in creation and use of data in the current times, there is need for efforts at all levels and across domains to ensure more responsible behaviour among its stakeholders particularly users in scientific work. But there seems to be less debate and concern towards need for desirable level of efforts in data citation space in India. It is evident by very less relevant content on this topic from any search being carried out in search engines about the discussion, debate and documents available in public domain. The advent of computing and communication networks like Internet has facilitated the availability and accessibility of kinds of data which earlier scientist would always wish to have strengthen and substantiate their arguments and findings. The limitations of access to data earlier mostly led to a give up situation among academic practitioner and later partly addressed by creation of synthetic data as a closer proxy for their teaching or testing needs due to advances in computing technology and applications. Now, we have both. We can craft closest simulations as well as access actual data. The liberal behaviour to share data across regions of the globe was quite skewed earlier, but one can see a gradual shift towards a normal one where more and more data creators and information owners are parting with data and making it accessible to world at large.

7 Sharing, Access and Availability

In academic space, data usage is sufficiently routine. Much of this usage is dependent on availability of shared data. Therefore, information sharing and information management are becoming more and more important in today's context. Though efforts are going on to make data sharing as part of the scholar responsibility, still there are challenges. Even who share data, do so in a manner that much of the desired information about the sources, process and other requisites fall short of needed information. This could be due to reasons like kind of technology application, format, access controls, accessibility infrastructure but many times due to apprehensions of the providers.

8 Data Sources

The use of data from secondary sources is growing at a rapid pace, but sensitivity towards citation of the sources has not evolved at the same pace. One of the major gaps in academic research particularly at the master level research work by the young scholars, who are proficient in finding and digging digital data from variety of archives on Internet, forgoes many times complete citation. The attitude issues towards data source citation and source attribution is not a localised concern but quite distributed across the globe. Applications capable of scanning such issues and other discrepancies in scientific work have widespread usage indicative of prevent attitudinal issues in scientific process. Data usage and attribution are a part of that.

9 Accountability

Desirable type of attitude and right levels of accountability in data sharing and attribution depend much on constitution of individualistic behaviour towards passion for global good and inclusion. But it is not easy to make it happen with the majority of the active audience in scientific endeavour who are mostly adults and old where habits are in a hardened form. Hence, it needs to be done at a much earlier stages like at the levels of school-going children or even earlier as a social policy intervention. There is a need to consider right reforms in the education policy at the primary and secondary levels. Reforms, which foster collaboration, respect for other's rights, value complementary additions in team set-ups, etc., can possibly usher voluntary engagement of scholars and scientists as contributors who want to see a larger good.

10 Research Gaps

A reasonable proposition in the current situation for some local scholars, interested in world of data science or research process or anything related to R&D development, is to identify and prioritise research gaps in this domain of data sharing and data citation. There is a need for basic descriptive studies to advance studies with scope across discipline, institutions and countries. Various scholars have investigated such issues of bilateral nature and outcomes of such studies are highly valued by scientists as well as policy makers.

11 DC Implementation

On the implementation side, data citation and best citation practices could be mandated through governance framework among institutions of all sorts till behavioural shifts take place which may take some time to fructify. To work on attitude, behavioural shifting models need to be tested well; otherwise, it can lead elsewhere. Better run pilots to build cases as it works in most cases including India. Once results of good practices start getting visible, shifts may happen faster. Indian culture, in most cases, is known for voluntary shifts once it recognises benefits of shifts in pilots or pockets.

12 Awareness Steer

There is also need for creating greater awareness and complete understanding of the requirements among the scientific community and weave the requisites for data citation in the guidelines of agencies who engage scientists in various forms like conferences, publications and any other forum of scholarly display. While there is evidence to justify reform-based mandated research practice among scholars and scientists, but there are also large-scale successes of behavioural shifts in India to show that there is plausibility of voluntary behavioural changes particularly collaboration. Cooperative concept has been an astounding success even among the farmers to share profits and business responsibilities. There is greater need to make contributors aware of the benefits of citation and handhold cases in all regions of the scientific community. Many times, new processes create an unnecessary scare unless these are tried out at least once. Many times, people do not do lot of good things not because they do not want to do it but because they do not know how to do it, however, simple it may be.

UK-India Research and Innovation Collaboration: Taking Forward Collaboration on Big Data and E-infrastructure

Nafees Meah

1 Introduction

We are living in an era of 'Big Data'.[1,2] That is to say data sets so large, complex and rapidly generating that they cannot be processed by traditional information and communications technologies. In 2012, the USA announced the National Big Data Research and Development Initiative to address the challenge and opportunity of 'Big Data'.[3] In the UK, the Big Data Institute at the Li Ka Shing Centre for Health Information and Discovery was launched by Prime Minister David Cameron in May 2013.[4] The Big Data Institute will focus on the analysis of large anonymised medical data sets—such as is collected through NHS electronic patient records, DNA sequencing, clinical trials and national registries—in an effort to improve detection, treatment and prevention of a range of conditions. More recently, in October 2015, the UK established Alan Turing Institute as its National Institute for Data Science.[5]

It is not only in the physical sciences where the significance of Big Data is becoming more and more evident. Recently in the UK, as part of its Digital Transformations in the Arts and Humanities Theme, the Arts and Humanities Research Council (AHRC) brought together academics and representatives from non-academic organisations that are interested in analysing, interpreting,

---

[1]http://www.mckinsey.com/insights/business_technology/big_data_the_next_frontier_for_innovation.
[2]http://www.cra.org/ccc/files/docs/init/bigdatawhitepaper.pdf.
[3]http://www.whitehouse.gov/blog/2013/04/18/unleashing-power-big-data.
[4]http://www.timeshighereducation.co.uk/news/oxford-big-data-centre-to-get-30-million/2003661. article.
[5]https://turing.ac.uk/.

N. Meah (✉)
RCUK India Office, New Delhi, India
e-mail: nafees.meah@rcuk.ac.uk

© Springer Nature Singapore Pte Ltd. 2018
U. M. Munshi and N. Verma (eds.), *Data Science Landscape*,
Studies in Big Data 38, https://doi.org/10.1007/978-981-10-7515-5_22

interrogating and visualising (etc.) Big Data from an Arts and Humanities perspective. AHRC will shortly be launching a funding call in this area.[6]

The UK's world-class social science research infrastructure is also a key factor in providing the vibrant research environment. Its suite of longitudinal studies is internationally renowned. The UK has been at the forefront of activities to improve and open up access to administrative and secure data. It has an international reputation for developing some of the most ground breaking methodological tools and techniques. Ensuring the exploitation and use of these data resources will allow the social sciences to play an important role in addressing the major global challenges.[7]

In India, the Unique Identification Authority of India has been charged with implementing a mammoth task of registering and assigning a unique 12-digit ID to every Indian resident—some 1.2 billion people—by 2020.[8] That aside, most of the Big Data Initiatives in India revolve around spatial planning, climate and weather prediction and monitoring, astrophysics and life sciences research.[9]

To store and analyse the data that has been and will be generated from these activities, India has embarked on a major programme to develop the necessary E-infrastructure and new national super computer facilities for data intensive research. The Planning Commission drew up a National Supercomputing Roadmap in its 12th Plan document (2012–2017). The roadmap projected development of high-capacity peta-scale supercomputers and technologies with a view to creating an opportunity for building exa-scale supercomputing capability.

The Centre for Mathematical Modelling and Computer Simulation in Bangalore has been repositioned as the nucleus of the Fourth Paradigm Institute. The institute aims to develop capacity and capability in the area of hardware and application software to meet the demands of the scientific community from all disciplines which are data intensive. The Biotechnology Information System Network (BTISnet) aim is to use the power of computation in biology, especially to handle massive data sets generated by new generation high-throughput technologies. The BioGrid India which was established by BTISnet is led by the Supercomputing Facility for Bioinformatics and Computational Biology, Indian Institute of Technology, New Delhi.

Since 2012, India has been connected to the Global Optical Ring Network for Advanced Applications Development (GLORIAD). This high-speed connectivity enables scientists in India to engage in advanced Big Data science research (such as atmospheric and oceanic sciences, biological sciences, high-energy physics and astrophysics, and material science) in partnership with global scientific community.[10]

---

[6]http://www.ahrc.ac.uk/News-and-Events/Events/Pages/Big-Data-Workshop.aspx.

[7]http://www.esrc.ac.uk/news-and-events/announcements/25683/Big_Data_Investment_Capital_funding_.aspx.

[8]http://uidai.gov.in/.

[9]http://12thplan.gov.in/.

[10]http://www.gloriad.org.

Sharing of infrastructure to support the global research community is becoming increasingly important.[11] For example, in the life sciences, the development of new technologies, for example, next-generation DNA sequencing, means that data produced is doubling every few months, and this rate continues to increase. In addition, researchers are constantly generating new types of data which needs to be managed. The collection, curation, storage, archiving, integration and deployment of all of these biomolecular data are a significant challenge and require international collaboration to be carried out successfully.

2 RCUK—India Research and Innovation Partnership

Research Councils UK (RCUK) is the name given to the strategic partnership of the UK's seven Research Councils who annually invest around £3 billion in research. It supports excellent research that has an impact on the growth, prosperity and well-being. RCUK fosters international collaborations and provides access to the best facilities and infrastructure around the world. To maximise the impact of research on economic growth and societal wellbeing, it works in partnership with other research funders including Innovate UK, business, government and charitable organisations. Over the past few years, RCUK has established a strong partnership with India.

Since 2008, RCUK and funding agencies of the Government of India have, together, developed a substantial portfolio of joint research and innovation programmes covering the themes: nuclear energy, renewable energy and energy access, food security and agritech, water and climate change, chronic disease and digital economy.[12]

In 2014, the UK and Indian Governments established a joint Task Force to identify potential *Grand Challenge* areas for future UK and India joint research and innovation. The Task Force met twice in 2014. Key research and innovation funders, and senior decision makers from both countries, attended the meetings and represented a wide spectrum of interests from knowledge to delivery on diverse themes. The Task Force identified three interdisciplinary, *Grand Societal Challenges* on:

- Sustainable cities and urbanisation
- Public health and well-being
- Energy–water–food nexus

It also identified two underpinning capabilities on:

- High value manufacturing
- Big Data.

---

[11]http://www.nerc.ac.uk/research/capability/environmental.asp.

[12]http://www.rcuk.ac.uk/RCUK-prod/assets/documents/india/ShapingTheFuture.pdf.

The areas identified by the UK-India Taskforce align closely with the ambitions of the Government of India (e.g. Smart Cities Mission, Digital India, Make in India, National Mission on Clean Ganga, etc.). In these areas, the RCUK-India research and innovation partnership has the potential to make a significant contribution in the future.

3   RCUK India Roundtable on Big Data and E-infrastructure

RCUK India, which represents all the Research Councils in India, organised a roundtable bringing together leading academics, policy makers and thought leaders from India and the UK.[13] The roundtable discussions considered four themes but also concluded that collaboration on Big Data between the two countries ought to focus on the Grand Challenge areas.

3.1   Theme 1—Application Domains for Big Data in Social Media, Industry, Government and Research Communities

Under this theme, it was concluded that there were opportunities for collaboration in:

1. Open innovation for affordable health care
2. Integration and consolidation of data sets
3. Big Data analytics
4. Smart sensor applications
5. Collaboration on Data Centres for Social Sciences

3.2   Theme 2—Legal and Security Frameworks (Cybersecurity)

Under this theme, it was concluded that there were opportunities for collaboration in:

1. Comparative review of legislation on data privacy
2. Understanding public perceptions of privacy
3. Procedural rules to govern access to Big Data sets
4. Data confidentiality and disclosure control

---

[13]http://www.rcuk.ac.uk/RCUK-prod/assets/documents/india/RoundtableBigData.pdf.

3.3 Theme 3—Sharing Methodologies and Developing Joint Approaches

Under this theme, it was concluded that there were opportunities for collaboration in:

1. Creating joint data hubs for different sectors (e.g. health)
2. Low-cost approaches to data visualisation
3. Education and training in data sciences
4. Machine learning
5. Low-cost methodologies for data collection

3.4 Theme 4—E-infrastructure and Harmonisation of Unstructured Data

Under this theme, it was concluded that there were opportunities for collaboration in:

1. Building human capital for Big Data analytics
2. Data management and data archiving methods and standardisation
3. Tools and techniques for working with domain experts
4. Underpinning communications systems for data delivery infrastructure
5. New algorithms and standards for cloud computing.

4 Taking Forward Research and Innovation Collaboration on Big Data and E-infrastructure

Having identified potential areas of collaboration, RCUK and a number of research funding agencies in India have been exploring routes to take these forward. Economic and Social Research Council (ESRC) is currently working with the Indian Council of Social Science Research (ICSSR) and the Ministry of Statistics and Programme Implementation on developing an Indian Data Centre for Social Sciences. Discussions are underway to develop a joint UK-India research programme on cybersecurity. The Natural Environment Research Council's (NERC) and the Ministry of Earth Sciences's (MoES) joint research programmes will be generating huge and varied data sets and discussions that are under way on optimising the management and handling these large volumes of variable data sets. RCUK and British Council are also in discussion on utilising funding available under the Newton Bhabha Programme to develop Ph.D. Partnerships and Ph.D. Placements linking academia and industry in both the UK and India[14] to work on Big Data issues.

However, the recent establishment of the Alan Turing Institute in the UK and the Department of Science and Technology's announcement of its new Big Data Initiative offers an opportunity to substantially enhance collaboration between the UK and India on Big Data and E-infrastructure.[15]

---

[14]https://www.britishcouncil.in/newton.
[15]http://dst.gov.in/big-data-initiative-1.

The Urgent Need to Overcome a Generation Gap in Public Data Management Practices for Digitalized India

Biplav Srivastava

1 Introduction

It is a non-brainer in today's knowledge-driven economy that data is king [1]. In this context, India has been collecting data for decades in areas as diverse as agriculture, weather, energy, rivers, health, and space. However, the data management practices are at least a generation behind where they need to be to match the global state-of-the-art and deliver much needed economic results. I spoke at a panel at an international big and open data workshop at New Delhi, India, in first week of November 2015 [2], and this note summarizes my take on what needs to change.

2 Generation Gap in Public Data Management

In order to bridge the gap in public data management, my perceptiveness suggests to factor some changes as indicated below.

First, consider the state-of-the-art. If one searchers for Lata Mangeshkar on BBC Music [3], the information is retrieved not just from BBC (tracks played) but also from encyclopedia's like MusicBrainz (playlist) and Wikipedia (biography) [4]. The information is comprehensive, useful and accurate, delivered fast.

Now, sample the questions a public officer should be interested to solve: Knowing diarrhea is a water-borne disease and that we have both water quality and disease data in the country, can we find out in which parts of the country diarrhea is caused due to river pollution? Further, knowing money is spent on river cleaning does it bring down cases of diarrhea in a specific region? Going a step ahead, is it

B. Srivastava (✉)
IBM Research, Yorktown Heights, NY, USA
e-mail: biplavs@us.ibm.com

© Springer Nature Singapore Pte Ltd. 2018
U. M. Munshi and N. Verma (eds.), *Data Science Landscape*,
Studies in Big Data 38, https://doi.org/10.1007/978-981-10-7515-5_23

more effective in reducing diarrhea deaths by spending the limited money available to treat immediate diarrhea cases or in tackling river pollution or which degree of both? These are just a slice of questions in public health, and hundreds more are asked everyday in all spheres of Indian economy.

Unfortunately, existing data is not in the form to be exploited and this failure of data access and analysis leads to costly, desperate, and ad hoc decisions like endemic unnecessary deaths due to dengue, malaria, and Japanese encephalitis; closing tanneries causing job-loss; banning truck movements causing price increases, and importing sand causing foreign currency wastage that could have been better tackled with better alternatives revealed by data.

The situation is slightly improving but a concerted effort is urgently needed to overcome what I call is a generation gap in data practices.

As a recap, India adopted the policy for sharing data called National Data Sharing and Accessibility Policy (NDSAP) in March 2012, and started implementing it via NIC later that year [5]. As of November 1, 2015, the site has over 18,000 resources from 100+ government organizations. Although commendable, the current investments and practices into public data collection and management technologies do not match up to the needs of digitalized India. The visualizations (in 100s) and apps (10s) using the published data are still low and only now ecosystem is developing to use these datasets productively via hackathons and competitions [6].

But there are also regressive trends. For example, I learnt that another organization, Indian Council of Social Sciences Research (ICSSR) will start publishing data from the government [6] under same policy but with a slight syntactical difference in process and very different data hosting platform! [7]. What the decision makers miss is that the value of data is dictated by how easy it is to use them. In fact, data has no value if it cannot be used. Having multiple data publishers implementing the same NDSAP policy would lead to wastage of precious public resources, confusion to data users, and hurdle to data usage, defeating their main objective of data usage in the first place.

Instead, following urgent steps are needed to drive data usage and quality by:

1. Semantic linking of datasets

No data is useful just by itself. Instead, they need to be inter-linked with others so that users can avoid redundancy and derive superior value. Tim Berners-Lee summarized it as a 5-star data quality guideline [8]. For example, Indian data can be improved by linking them with the official list of districts and states in the country, a vocabulary for which is already on data.gov.in. Today, Indian data is at 2–3 stars and many will move to 4–5 stars with just this simple addition.

2. Organizing data by case studies and analytics

Data has to be discoverable to be of value. Since the problems in India are perennial and nature of solutions sought is also well known, all we need to do is formalize them as case studies and use them to organize data. Common examples are environment where one would want to know pollutant levels, health where one would want to know disease estimates and availability of health facilities, agriculture where one would want to know crop yields and diseases, and crime where one would want to know about prevalent crimes and actions taken. Furthermore, if we know error rates of available analytics, this can help drive new innovations and competitions to improve them [6].

3. Promoting a culture of data sharing and use

Only a fraction of data generated is available in public. This can change if everyone, no matter how big or small, individual organization or government body, publishes the data they generate to the extent possible. A common example is temperature of a place, which, if available for all intersections in a city, can be valuable to predict air pollution of that city. Further, data justification should be sought for any public decision being taken. Data publishing and usage will raise awareness about the ease and benefits of sharing data.

3 Concluding Remarks

In summary, data has enormous potential to revitalize economy and tackle societal issues. However, our current focus of piece-meal data publishing is insufficient. We need to improve their quality with semantic linking, increase their discoverability via case studies and analytics, and promote a culture of data sharing by everyone.

Biplav has been engaged in innovative technologies to enable people to make data-driven decisions for two decades. He has worked extensively with open data globally. This is a personal opinion.

References

1. McKinsey Open Data Estimate. At: http://e-pluribusunum.org/2013/10/29/mckinsey-estimates-open-data-could-add-more-than-3-trillion-in-economic-value/, 2013.
2. Workshop on "Big and Open Data: Evolving Data Science Standards and Citation Attribution Practices". At: http://www.insaindia.org/pdf/Final-Brochure-6-oct-2015.pdf. November 2015.
3. Lata Mangeshkar on BBC Music. At: http://www.bbc.co.uk/music/artists/aeb71bd8-447d-4415-8ea1-2b7d664f67e1. Accessed November 2015.
4. BBC Music FAQ. At: http://www.bbc.co.uk/music/faqs#where_does_the_data_come_from. Accessed November 2015.
5. Indian open data. At: http://data.gov.in, November 2015.
6. DataView 2016 competition. At: https://www.facebook.com/dataview2016, November 2015.
7. Indian Council of Social Sciences Research (ICSSR) open data. At: http://14.139.116.25/icssr_dr/index.php, November 2015.
8. 5 star data rating. At: http://5stardata.info/en/. November 2015.

Data Security in Cloud-Based Applications

Surabhi Pandey, G. N. Purohit and Usha Mujoo Munshi

1 Introduction

Global business challenges and expeditious growth of internet services especially in the last decade have driven business organizations to seek an emerging technology for doing business. Cloud computing business application (SaaS) is gaining momentum across global business markets. In context to our country, most organizations are increasingly realizing the underlying benefits of this IT service delivery model, this has generated interest in cloud environment and its adoption among sundry industries. However, it is desirable to consider and analyze the significant points of existing facts and substantive information about various factors influencing both adoption and denial of cloud-based business IT solutions in several business organizations.

In cloud services, customers' data hosted on third party servers in the same database, so security is always a concern, though service providers deliver best of their services to secure and maintain confidentiality of customer's business sensitive data despite there is chance of human error and authorization [1].

S. Pandey (✉) · U. M. Munshi
Indian Institute of Public Administration, New Delhi, India
e-mail: dr.pandeysurabhi@gmail.com

U. M. Munshi
e-mail: umunshi@gmail.com

G. N. Purohit
Computer Science, Banasthali University, Banasthali, Rajasthan, India

© Springer Nature Singapore Pte Ltd. 2018
U. M. Munshi and N. Verma (eds.), *Data Science Landscape*,
Studies in Big Data 38, https://doi.org/10.1007/978-981-10-7515-5_24

321

2 Data Security

There are certain technical challenges involved in the adoption of cloud-based technology and one of the biggest challenges is data security [2]. Data security is paramount task for application and data architects while designing cloud-based business application architectures. Therefore, the foremost priority of cloud application providers is to design and develop robust application and multi-tenant efficient data architecture to ensure that each customer can access their own data only.

Data security challenges [3] related to cloud computing can be categorized into four broad categories:

1. Authentication: Restricting access to data as per user role and authorization.
2. Data protection: Data security both in transit and at rest.
3. Disaster recovery and contingency planning for possible data security breach.
4. Trust: Trust on service providers and hosting of data on third party.

Since data security is put on highest priority by cloud service providers [4, 5], it has become most important task for any SaaS application/data architect. Building and enhancing data security requires detailed analysis of different levels of application and identification of the areas where risks are more with the solution to address them promptly [6].

There are following three basic security patterns to provide right type of data security at right areas.

- **Filtering**: Filtering isolates different customer data from each other and acts like filter by introducing intermediate layer. It makes it appear for customer that only data pertaining to that customer is stored in the database.
- **Permissions**: Allows only authorized users can access data in the application by implementing access control lists (ACLs).
- **Encryption**: Advance data encryption is applied to secure tenant's business-critical data from unauthorized access.

3 Solutions for Data Security in Cloud Computing

- Data security is a main obstacle in adoption of different cloud computing services including SaaS. Following is few data security measures that can be implemented to enhance data security in cloud solutions [7].
- Network security: A user can reject to access of Internet/Web-based service by using IP spoofing and phishing technique which may be responsible for security threat. This issue can be resolved by digital signature technique. Secure socket layer (SSL) protocol is primarily used for making communication channel more

secure and handling message transmission extra secure on the Internet. Hence, hacking can be avoided.

- Encryption algorithm: A cloud service provider uses cryptography technique to encrypt the user's information by strong encryption algorithm [8]. But there are chances that that encryption fate can make data totally inoperative, and encryption also complicates the availability of data. To resolve this problem, the cloud provider must offer proposal that encryption algorithm was designed and tested by experienced IT cryptography specialists.
- Investigation support: There are certain audit tools available which facilitate the users to control and regulate how their data is used, stored, protected and verify policy execution. But investigation of illegal activity is quite difficult because data for multiple customers may be collocated and may also be geographically spread across set of hosts and data centers. To solve this problem, audit tools must be contractually committed beside with the proof.
- Backup: An important data may lose due to natural disaster, which may also damage the physical devices. To resolve this problem, backup and recovery of information are the key to guarantee of data and service provided by vendor.
- Customer satisfaction: It is really difficult for the customer to verify the currently implemented security measures, techniques, and initiatives of a cloud computing services provided by the service provider because the client usually has no access.

4 Data Security with Encryption Techniques

In order to enhance data security in cloud applications, an optimum algorithm using existing symmetric AES RInjndael 256 bits is proposed to encrypt data in database [9].

Existing AES encryption and decryption has many applications [8]. It is used in cases where data is too sensitive that only the authorized people are supposed to know and not to the rest, such as secure communication—smart cards, RFID, ATM networks, image encryption, secure storage, etc.

4.1 AES Algorithm Specification[1]

The advanced encryption standard *(AES) is a symmetric-key block cipher* proven by the US National Institute of Standards and Technology (NIST) in 2001.

---

[1]Note: AES Specification from: [13]

Table 1 Key-block-round combinations [9]

Bit pattern	Key length (Nk words)	Block size (NB words)	No of rounds (NR words)
AES 128	4	4	10
AES 192	6	4	12
AES 256	8	4	14

The algorithm described by AES is a symmetric-key algorithm [10], meaning the same key is used for both encrypting and decrypting the data. AES has defined three versions, with 10, 12, and 14 rounds. Each version uses a different cipher key size (128, 192, or 256), but the round keys are always 128 bits. Rijndael is an iterated block cipher [11] with a variable block length and a variable key length. The block length and the key length can be independently specified to 128, 192, or 256 bits (Table 1).

5 Implementation

An implementation of the AES algorithm shall support at least one of the three key lengths: 128, 192, or 256 bits (i.e., Nk = 4, 6, or 8, respectively). The key length is represented by Nk = 4, 6, 8. The number of rounds is represented by Nr, performed during the execution of the algorithm is dependent on the key size.

When Nr = 10, then Nk = 4,
When Nr = 12, then Nk = 6,
When Nr = 14, then Nk = 8.

For both its cipher and inverse cipher, the AES algorithm uses a round function that is composed of four different byte-oriented transformations:

- Byte substitution using a substitution table (S-box),
- Shifting rows of the state array by different offsets,
- Mixing the data within each column of the state array, and
- Adding a Round Key to the State.

5.1 *Encryption* of Database Value: Symmetric AES RInjndael 256 Bits

Step 1: Initialize
static readonly string PasswordHash = "P@@Sw0rd";
static readonly string SaltKey = "S@LT&KEY";
static readonly string VIKey = "@1B2c3D4e5F6g7H8";

Step 2: Apply encrypt function on string that need to be encrypted (email) Using symmetric key.
Step 3: Invoke function encrypt (string)
Step 4: Convert string into byte & store in variable
Step 5: Encrypt the string using Rijndael(AES method that use password,salt-key&vikey for encryption
//keyBytes = newRfc2898DeriveBytes(PasswordHash,Encoding.ASCII.GetBytes (SaltKey)).GetBytes(256/8);
symmetricKey = new RijndaelManaged() {Mode = CipherMode.CBC, Padding = PaddingMode.Zeros};
encryptor = symmetricKey.CreateEncryptor(keyBytes, Encoding.ASCII.GetBytes (VIKey));//
Step 6: Store in database
Step 7: Exit

5.2 Decryption of Database Value: Symmetric AES RInjndael 256 Bits

Steps involved in decryption of database value:-

Step1: Get the encrypted (column)
Invoke Decrypt () function.
Step2: Convert encrypted string into bytes
//byte[] cipherTextBytes = Convert.FromBase64String(encryptedText);
byte[]keyBytes = newRfc2898DeriveBytes(PasswordHash,Encoding.ASCII. GetBytes(SaltKey)).GetBytes(256/8);
varsymmetricKey = new RijndaelManaged() {Mode = CipherMode.CBC, Padding = PaddingMode.None};//
Step3: Decrypt the string using same Rijnadael(AES) key same as in encryption.//
vardecryptor = symmetricKey.CreateDecryptor(keyBytes, Encoding.ASCII.GetBytes (VIKey));////return Encoding.UTF8.GetString(plainTextBytes, 0, decryptedByteCount). TrimEnd("\0".ToCharArray());//Step4: Display data
Step5: Exit

6 Conclusion

Data security and risk of loss of business-critical data are major concerns in cloud computing as single centralized server is serving multiple customers. Considering all cloud computing security challenges, it is also the duty of the customer to check and verify whether the cloud application service provider has implemented all essential data security measures to secure the customer's data from unauthorized access. Before finalizing the cloud business application service provider, an independent inclusive data security assessment is highly recommended by neutral data security consultants.

References

1. Gellman, R. (2009). *Risk to privacy and confidentiality from cloud computing. World Privacy Forum, 1*(1), 1–26. Retrieved from www.worldprivacyforum.org.
2. Sun, D., Chang, G., Sun, L., & Wang, X. (2011). Surveying and analyzing security, privacy and trust issues in cloud computing environments. *Procedia Engineering, 15,* 2852–2856.
3. Mather, T., Kumaraswamy, S., & Latif, S. (2009). Cloud security and privacy an enterprise perspective textbook. O'Reilly Media Inc.
4. Katzan, H. (2010). Identity and privacy services. *Journal of Service Science, 3*(2), 1–13.
5. Katzan, H. (2010). On the privacy of cloud computing. *International Journal of Management and Information Systems, 14*(2), 1–15.
6. Rittinghouse, J., & Ransome, J. (2010). *Cloud computing: implementation, management, and security.* Boca Raton, FL: CRC Press.
7. Gonzalez, N., Miers, C., Redígolo, F., Simplício, M., Carvalho, T., Näslund, M., & Pourzandi, M. (2011). A quantitative analysis of current security concerns and solutions for cloud computing. *Journal of Cloud Computing: Advances, Systems and Applications.* New York: Springer, online. ISSN: 2192-113X.
8. Stallings, W. (2011). Cryptography and network security principles and practice 5e. 2011, 2006. Publishing as Prentince Hall. ISBN 10: 0-13-609704-9, ISBN 13: 978-0-13-609704-4.
9. Pitchaiah, M., & Daniel, P. (2012). Implementation of advanced encryption standard algorithm. *Research paper published International Journal of Scientific and Engineering Research, 3*(3), March 1, 2012. ISSN 2229-5518.
10. Foruzan Behrouz, A., & Mukhopadhyay, D. (2010). Cryptography and network security 2e. Book published by Tata Mac Graw Hill Education Pvt.ltd. ISBN 13: 978-0-07-070208-0. ISBN 10: 0-07-070208-X.
11. Daemen, J., & Rijmen, V. (1820). The block cipher Rijndael smart card research and applications, LNCS 1820 (pp. 288–296). New York: Springer.
12. Habib, S. M., Hauke, S., Ries, S., & Muhlhauser, M. (2012). Trust as a facilitator in cloud computing: a survey. Journal of cloud Computing. New York: Springer, online. ISSN: 2192-113X.
13. Rong, C., Nguyen, S. T., & Jaatun, M. G. (2013). Beyond lightning: A survey on security challenges in cloud computing. *Published: Computers and Electrical Engineering on science direct, 39*(1), 47–54. January 2013.
14. Specification for the Advanced Encryption Standard (AES),∥ Federal Information Processing Standards Publication 197, November 2001.
15. URL Referred: https://crackstation.net/hashing-security.htm/

Author Index

© Springer Nature Singapore Pte Ltd. 2018
U. M. Munshi and N. Verma (eds.), *Data Science Landscape*,
Studies in Big Data 38, https://doi.org/10.1007/978-981-10-7515-5

Subject Index

© Springer Nature Singapore Pte Ltd. 2018
U. M. Munshi and N. Verma (eds.), *Data Science Landscape*,
Studies in Big Data 38, https://doi.org/10.1007/978-981-10-7515-5

Printed in the United States
by Book masters

Printed in the United States
By Bookmasters